Dominic Couzens
Artwork by David Nurney

BIRDS ID INSIGHTS

BLOOMSBURY WILDLIFE
Bloomsbury Publishing Plc
50 Bedford Square, London, WC1B 3DP, UK

BLOOMSBURY, BLOOMSBURY WILDLIFE and the Diana logo are trademarks of
Bloomsbury Publishing Plc

First published in 2013 in the United Kingdom as by New Holland Publishers Ltd
This edition published 2014

A catalogue record for this book is available from the British Library

Library of Congress Cataloguing-in-Publication data has been applied for

ISBN: HB: 978-1-4729-8213-1; ePub: 978-1-4729-1121-6; ePDF: 978-1-4729-1995-3

2 4 6 8 10 9 7 5 3 1

Printed and bound in India by Replika Press

Dominic Couzens
Artwork by David Nurney

BIRDS ID INSIGHTS

Identifying the More Difficult Birds of Britain

BLOOMSBURY WILDLIFE
LONDON • OXFORD • NEW YORK • NEW DELHI • SYDNEY

CONTENTS

FOREWORD

I know many birders who wake up sweating during the middle of the night, worrying about their ability to identify birds. This concern is often borne from peer group pressure and from the perceived impression that to be a proficient birder you have to be up with the latest scientific papers and have read all the field guides from cover to cover. I too used to be part of that neurotic band until the day I realised that birding didn't have to be so pedantic. It is just as possible to identify species without feather-by-feather analysis.

It must be said that there is nothing wrong with making identification mistakes. All of us do it, even the so-called experts. The more mistakes we make the better we become as birders. We have to learn to see birds as animals with personality and variable plumage patterns – not just as one-dimensional objects that have to be perfectly matched with an illustration in a book. That is the thing, so many of us ignore the obvious to concentrate on the minutiae of plumage detail. That is what so great about this book. Dominic Couzens and David Nurney have dispensed with the technical handbook approach and instead have settled for a more realistic, organic and refreshing way of identifying birds. It still is a very detailed book but in a very user-friendly format based around David's superb artwork and Dominic's concise yet illustrative text. Let's face it, most of us are not 'experts' and although we want to learn more about how to work out our Willow Warblers from Chiffchaffs, we also don't want the pressure of having to study notes on wing formulae for months on end.

Birds ID Insights will debunk your misconceptions of how hard some bird species are to separate and give you a great sense of confidence to get out there and enjoy the birds that you encounter. This is a book for people that absolutely love birds.

David Lindo, The Urban Birder

INTRODUCTION

This is a field guide to some of the trickier bird species of north-west Europe. It gives insights into identifying the birds by presenting a large number of detailed, labelled, quick-reference illustrations in an attractive and easy-to-compare style. There are usually more illustrations of a species than are available in most conventional field guides, and for some species there are whole 'galleries' that depict a wide range of plumages covering a whole year, or whole life of a bird. Furthermore, some groups of birds are depicted in a non-standard field-guide way, for example with different attitudes or shapes in flight, and with extensive direct comparisons to similar species. All in all, we have endeavoured to present a range of bird species in a refreshing way that will hopefully help the penny to drop for some difficult to identify puzzles.

The style of this book arose out of an ongoing series written by us every month within the pages of *Bird Watching* magazine. Our target audience is mainly 'improvers' who want a little more than a normal field guide has to offer. A majority of the illustrations presented here has been published in the magazine.

With the original articles aimed at a British audience, there was always a strong bias towards north-west European species, and that remains. However, we have added a good number of extra species, so the region covered by the guide includes France, Germany, Belgium, the Netherlands, Luxembourg, Denmark and the southern half of Scandinavia. Birds that are mainly confined to Iceland and Arctic Scandinavia are not included. We have also been sparing about including too many birds from the Mediterranean region, since this would have added a lot of extra species. The book should, however, be useful in many European countries outside the immediate area.

This book is not designed to replace a 'normal' field guide, but to complement it. We have not included easily identified and distinctive species, and for the most part have left out rarities. Instead the sections include the most difficult groups and most variable species – those that give the ordinary birdwatcher the most headaches most often. If the book can sort out some puzzles in the field for the birder, it will have fulfilled its purpose.

Layout

The organization of the book is simple enough. Each section has a very brief introduction detailing the species covered, together with their distribution, size (L=length; WS=wingspan), habitat and, if appropriate, providing a few general notes. This is followed by the annotated illustrations that are the core of the book.

A glossary of words and terms used in the book can be found on pages 266-7. It would be hard to use the book without some knowledge of the terminology of bird feathering and plumages. It is thus essential that the user learn the parts of the bird, as presented here, as the terminology is used throughout the book.

Bird Topography

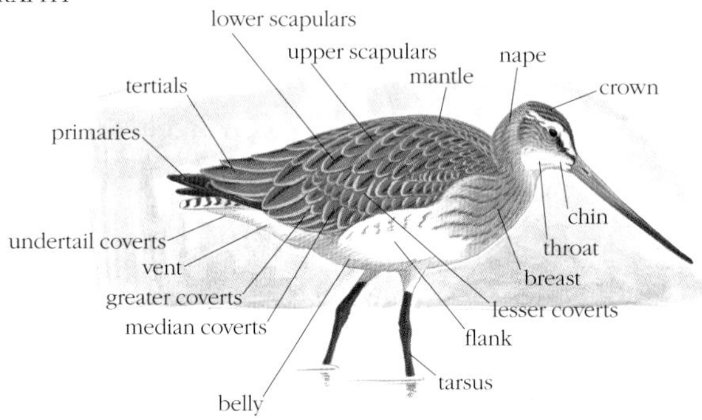

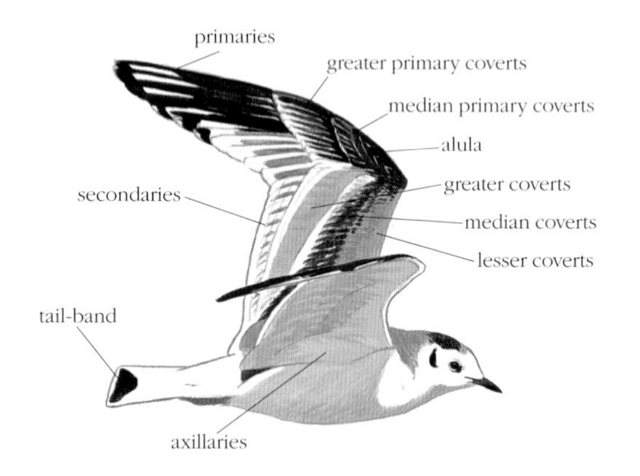

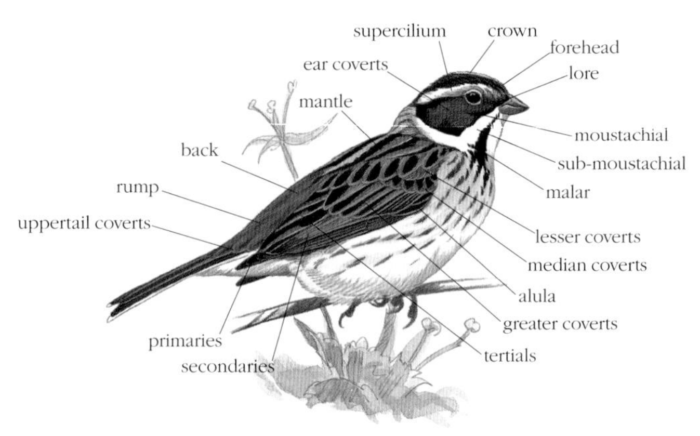

Plumage terminology

In this book we have been conventional in using 'summer' and 'winter' plumage, though 'breeding' and 'non-breeding', or even 'alternate' and 'basic', could have been used instead. For simplicity, summer plumage refers to the appearance of a bird when it is breeding, winter plumage to that exhibited the rest of the time. A few ducks have a so-called 'eclipse' plumage, which is exhibited for a brief period in mid-summer when birds are moulting, and males tend to resemble females until the autumn. On occasions we simply label a plumage according to the month or season in which it occurs. Where no age or sex is given an image refers to 'adult' plumage.

As birds mature they often acquire a set of transitional plumages to adulthood. Any youthful plumage exhibited before maturity may be termed, not surprisingly, 'immature', which is a useful blanket term for a host of confusing plumages (especially in gulls).

To be more specific, however, a young bird's first plumage when it has left the nest and grown feathers is termed 'juvenile'. This often changes, either by moult or wear, into a different appearance later in the autumn – a set of feathers that is worn during a bird's first winter of life, hence the term first winter. If the bird remains different from an adult into the next few months or years, the terminology describing its plumage will be as in the table below. Few birds take as long as three years to mature.

Another interesting variation in plumage arises not through growing maturity but by the existence of consistent colour patterns within a population (polymorphism). This is seen, for example, in skuas, which have dark, intermediate and pale morphs.

Bird names

Full English and Latin names are used in each section introduction, for example Northern Goshawk (*Accipiter gentilis*) and Black-legged Kittiwake (*Rissa tridactyla*). For reasons of space these are often shortened for the image captions where there is no risk of confusion, for example to Goshawk and Kittiwake.

Table of bird plumages

	J	F	M	A	M	J	J	A	S	O	N	D	Cal Yr
Plumage				Egg	Chick	Juv	Juv	Juv	1W	1W	1W	1W	1st
Plumage	1W	1W	1S	1S	1S	1S	1S	2W	2W	2W	2W	2W	2nd
Plumage	2W	2W	2S	2S	2S	2S	2S	3W	3W	3W	3W	3W	3rd
Plumage	3W	3W	3S	3S	3S	3S	3S	AW	AW	AW	AW	AW	4th
Plumage	AW	AW	AS	AS	AS	AS	AS	AW	AW	AW	AW	AW	5th

Key

Juv Juvenile
1W 1st winter
1S 1st summer
2W 2nd winter
2S 2nd summer
3W 3rd winter
3S 3rd summer
AW Adult winter
AS Adult summer

Whooper & Bewick's Swans

INTRODUCTION Mute Swan (*Cygnus olor*) [L 153cm] widespread resident. Whooper Swan (*Cygnus cygnus*) [L 153cm] breeds in Iceland and Scandinavia; winters widely further south. Bewick's Swan (*Cygnus bewickii*) [L 122cm] scarce and localized winter visitor, breeding in Arctic Russia.

Bewick's Swan (adult)

Whooper Swan (adult)
• Breast tends to bulge more than Bewick's
• Longer bill than Bewick's
• Evenly sloping forehead down to bill
• Longer, flatter body than in Bewick's, with more bulging breast
• Typically more yellow than black on bill
• Flat crown and wedge-shaped bill/head distinctive
• Often holds head pointing down

Swan bill patterns

Whooper Swan (adult)

- Bill considerably longer than in Bewick's
- Yellow goes past nostril on bill
- Yellow tends to make sharp angle to black, creating 'wedge' of yellow

Bewick's Swan (typical adult)

- Yellow on bill abuts to black making round or squared-off pattern, rather than deep wedge of Whooper

Bewick's Swan (adult, variation)

- Variable design of black and yellow on Bewick's bill

Bewick's Swan (juvenile)

- Pattern is shadow of adult bill, with pink where most of black will be, and yellow much paler
- Darker head than Whooper's

Whooper Swan (juvenile)

- As in Bewick's, bill pattern is shadow of adult's
- Paler plumage than that of other swans, so much so that juveniles can be hard to tell from white adults at distance

Mute Swan (juvenile)

- Easy to distinguish from other juveniles by black border between bill and head, and black line to eye
- Hint of black knob
- Tends to have slightly darker pink-brown plumage than Bewick's or Whooper juveniles

Grey Geese

INTRODUCTION Greylag Goose (*Anser anser*) [L 79cm] widely introduced; wild populations breed in Scotland, Iceland and Scandinavia and winter in Britain, Ireland and the Low Countries. Pink-footed Goose (*Anser brachyrhynchus*) [L 71cm] breeds Iceland, Greenland; winters locally in Britain and the Low Countries. White-fronted Goose (*Anser albifrons*) [L 71–74cm] breeds on tundra of Greenland and Arctic Russia; winters mainly along the North Sea coast. Taiga Bean Goose (*Anser fabalis*) [L 83cm] breeds in Scandinavian taiga belt in forest clearings; winters on the North Sea Coast and western Baltic; a few in eastern England and southern Scotland. Tundra Bean Goose (*Anser serrirostris*) [L 75cm] breeds on Russian tundra; winters patchily in western Europe. Canada Goose (*Branta canadensis*) [L 90–105cm] widespread introduced resident.

Separating grey and black geese

Canada Goose

- Bold patterning on head and neck (c.f. grey geese)
- Dark bill

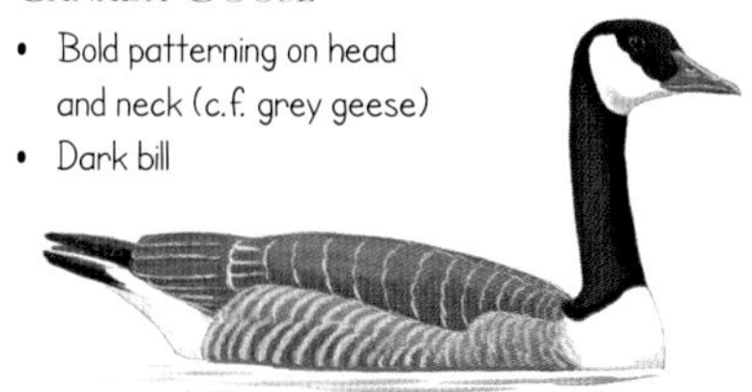

Greylag Goose

- Overall dull grey-brown ('grey goose')
- 'Wet-combed' effect on neck
- Colourful bill (c.f. 'black geese' such as Canada, Brent)

Grey goose silhouettes

Tundra Bean Goose

- Smaller than typical Taiga Bean (Tundra sized=undersized)
- Shorter neck than in Taiga Bean, but thicker than in Pink-footed
- Shorter bill than Taiga Bean's
- Different head profile from Pink-footed: typically wedge shaped, as in Taiga Bean

Greylag Goose

- Large head with huge long, triangular bill, quite distinct in shape compared with other goose bills
- Thick neck

Taiga Bean Goose

- As big as Greylag, but slimmer
- Long, narrow-based bill gives head wedge shape

Pink-footed Goose

- Small and compact
- Rounded head

Grey geese in flight

Pink-footed Goose

- Bill short, triangular
- Forewing pale grey
- Head and neck very dark, making sharp contrast with paler breast (more obvious than in Bean)

Tundra Bean Goose

- Smaller than Taiga Bean; about size of Pink-footed
- Shorter neck than Taiga Bean's
- Dark upperwing (c.f. Pink-footed)

Taiga Bean Goose

- Very large (as Greylag)
- Head very dark (as Pink-footed)
- Long, narrow neck may recall that of swan
- Underside of wing entirely dark (c.f. Greylag)

White-fronted Goose

- Large white patch at base of bill
- Bill pinkish (yellow-orange in Greenland race), without black tip (c.f. Bean, Pink-footed)
- Diagnostic earthy-brown, blotchy bars across lower breast and belly
- Distinctive square-shaped head
- Pale orange legs

Greylag Goose

- Long but thick neck - may look 'pinched in' mid-way along
- Legs pink (duller than Pink-footed)
- Diagnostically, underwing two-toned

Grey goose upperwings

Greylag Goose

- Upperwing predominantly pale ash-grey (paler than in Pink-footed)
- Brown 'square' leaks on to wing from mantle
- Back brown, contrasting with colour of most of wings (c.f. Pink-footed)
- Very pale grey rump, paler even than in Pink-footed

White-fronted Goose

- Dark brown upperwing, though not as dark as in Bean
- No contrast between colour of back and inner wing
- Broad, dark tail-band
- Narrow white tips to tail

Bean Goose (both species)

- Generally dark upperwing
- Very obvious white bars on mid-wing
- Dark tail-band and narrow white tip to tail
- Pale tips to tertials, useful for distinguishing species from dark-backed White-fronted

Pink-footed Goose

- Generally smoky-grey upperwing and back (little or no contrast)
- Grey rump
- Narrow pale tail-band

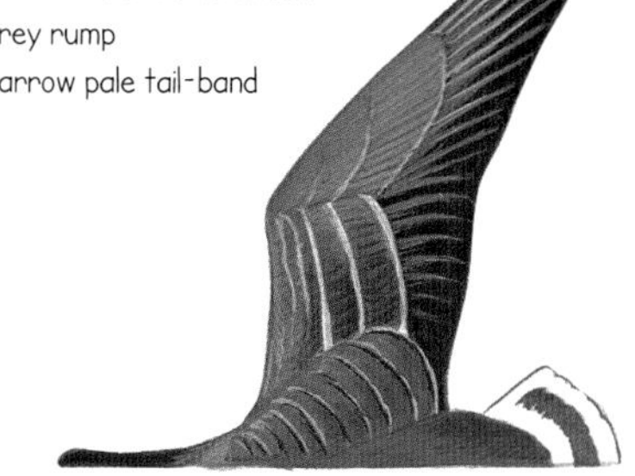

Bean and Pink-footed Geese

Taiga Bean Goose

- Long bill comparatively narrow at base (c.f. Greylag)
- Bill mainly orange (c.f. Pink-footed, Greylag); colour shared with White-fronted
- Bill usually has at least some black (c.f. White-fronted), but this varies
- Legs orange (c.f. Pink-footed)

Tundra Bean Goose

- Bill shorter than in Taiga Bean, but head still wedge shaped, with flat crown (c.f. Pink-footed)
- Bill black with small orange patch

Pink-footed Goose

- Bill mainly pink (c.f. Bean Goose) with black tip
- Legs bright pink
- Flank darker than upperwing (not so in Bean Goose)

FEMALE DABBLING DUCKS

INTRODUCTION Mallard (*Anas platyrhynchos*) [L 56cm] familiar and abundant resident throughout area; almost any waterbody. Gadwall (*Anas strepera*) [L 51cm] locally common on lowland freshwater lakes and marshes; more widespread in winter. Northern Pintail (*Anas acuta*) [L 56cm (not including male's extended tail-feathers)] localized breeding bird on larger marshes and meadows, mainly Scandinavia; common in winter, also on estuaries. Northern Shoveler (*Anas clypeata*) [L 49cm] widespread on well-vegetated freshwater lakes and marshes; common in same habitats in winter. Eurasian Wigeon (*Anas penelope*) [L 48cm] common breeder in Scandinavia and northern Britain in marshes and by lakes; widespread in winter, varied habitats. Common Shelduck (*Tadorna tadorna*) [L 60cm] common breeding bird, mainly around coasts.

Mallard

- Big, stocky duck with big, long bill
- Flattish crown sweeps down to make continuous curve on to bill
- Bill orange and brown, in rather untidy pattern (c.f. Gadwall); usually band of orange right across top of bill just before tip, lacking in Gadwall
- White sides to tail (c.f. Gadwall)

Pintail

- Pointed tail always conspicuous, even though nowhere near as spectacular as male's
- Slate-grey bill diagnostic (c.f. Mallard, Gadwall)
- Plain head lacks stripes, giving blank expression; head only strongly gingery in colour
- Overall, paler in colour than Mallard
- Plumage more delicately and intricately streaked and scalloped than in Mallard or Gadwall
- Breast paler than flanks (good feature at distance)

Gadwall

- Bill dark with orange sides, with two being divided along neat line
- Steep forehead and squarer head than Mallard's
- Whitish chin (c.f. Mallard)
- Dark tertials (c.f. Mallard)

Wigeon

- Rounded head with high crown
- Small, short, blue-grey bill
- Often noticeably dark around eye, as if it were wearing shades

Wigeon

- Completely different colour and pattern from all other female ducks: darker, reddish-brown, without streaks and speckles
- Individuals vary in colour - some (like this bird), have greyer heads than others; some are overall paler in colour than rich red-brown

Shoveler

- Diagnostic outsize bill
- Routinely swims with bill touching water's surface, as if glued to it
- High rear end
- Rusty tinge to plumage

Teal

- Small and short necked
- Greyer-brown than Mallard, Gadwall and Shoveler
- Green speculum often shows when swimming
- Distinctive white streak just below tail
- Face rather featureless

Gadwall (male)

- Gadwall sports most 'female-like' male plumage; often overlooked for apparent lack of colour
- Coal-black bill distinctive
- Black rear end contrasts with greyish lower belly (c.f. Wigeon, Shoveler)

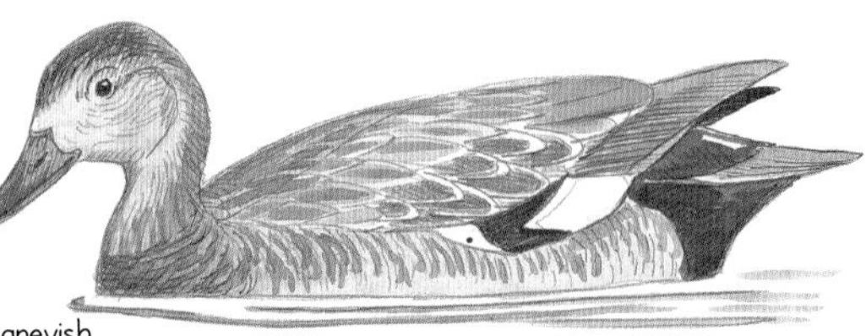

Juvenile Shelduck

- Resembles an adult which 'has been through a washing machine too many times'. Only the secondaries are colourful
- White face
- Pale eye-ring, contrasting sharply with dark head and neck
- Long pink bill
- Long legs different to other ducks

Dabbling duck feeding methods

Wigeon (female, grazing)

- Natural grazer, spending much time out of water, walking on grass in close-knit, medium-sized flocks
- Very prominent white belly contrasting with rich dark colour of neck and flanks
- Head subtly peppered with tiny dark spots

Gadwall (female, dabbling)

- Dabbling is placing bill flush to water's surface and filtering out food

Mallard (male, upending)

- Many dabbling ducks upend to feed at bottom of shallow water
- Curly tail diagnostic
- Bold black stripe down back

Female dabbling ducks in flight

Mallard

- Pale brown belly, paler than breast but lacking sharp contrast of Gadwall
- Wingbeats powerful and make loud swish, but not particularly fast
- White trailing edge to wing

Shoveler

- In contrast to the other ducks, flies off with distinctive drumming wingbeats
- Large bill
- Pot-bellied shape

Gadwall

- Very distinctive white belly
- Slightly faster wingbeats than Mallard's
- Narrower wings than Mallard's

Pintail

- Striking combination of long neck and small head
- White trailing edge to wing
- Tapered front and back
- Sharply pointed tail

Wigeon

- Prominent white belly with sharp contrast to rest of underparts
- Distinctly pointed tail

Garganey

- Larger and longer than Teal
- Slightly paler belly than Teal's
- Narrower white wing-bar than Teal's

Teal

- Very small and compact
- Seldom flies high
- Tendency to form tightly packed and well-coordinated flocks that may wheel around, twisting and turning
- Very fast wingbeats - so fast that unlikely to be confused with Mallard

Dabbling duck flight silhouettes

Pintail

- Profile distinctive, with very long neck and pointed tail
- Narrow, pointed wings
- Rather small head
- Flies with fast wingbeats and often travels high, seldom in large flocks (20 would be a lot)

Mallard

- Large, well-proportioned duck
- Long, fairly thick neck
- Slower wingbeats than those of other ducks

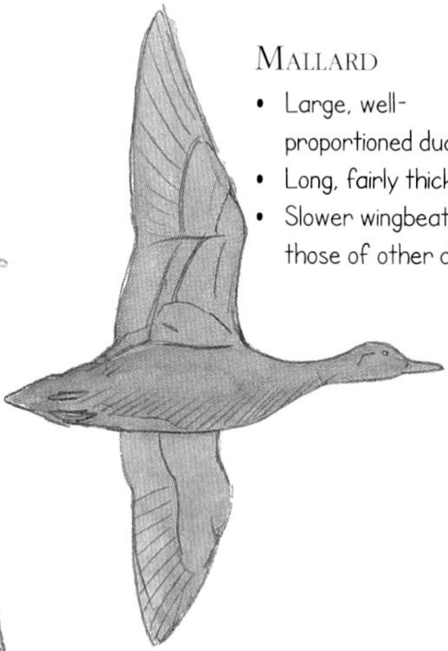

Shoveler

- Short, thin neck
- Wings appear set well back on body
- Tendency to fly high (as Pintail)
- Seldom in large flocks; usually only in single figures

Gadwall

- Shorter and narrower wings than Mallard's
- Smaller head and bill than Mallard's
- Usually in small groups, but joins flocks of other species.

Wigeon

- Short neck and bulbous head
- Pointed tail
- Often in large flocks (50 is typical)
- Like Pintail, often flies high

FEMALE TEAL & GARGANEY

INTRODUCTION Common Teal (*Anas crecca*) [L 35cm] widespread and common throughout area on freshwater pools and smaller lakes, often with fringing vegetation; also reservoirs and estuaries in winter. Garganey (*Anas querquedula*) [L 38cm] summer visitor March–October, marshes and shallow freshwater lakes; uncommon but more widespread on migration.

Garganey

- Distinctly longer bodied than Teal
- More substantial rear end, high at back
- Darker tertials than in Teal, with more obvious white fringes
- Flank feathers have white fringes only (double fringe in Teal)
- Flatter crown than in Teal
- Big, long grey bill
- Darker crown than in Teal
- Broad supercilium, whiter than Teal's
- White spot at base of bill often obvious

Teal

- Green speculum often shows while swimming (lacking in Garganey)
- Distinctive white streak just below tail
- Subtle spotting below tail
- Much plainer face than Garganey's
- Dark eye-stripe, but less obvious than in Garganey
- Often pink/orange base to bill (not in Garganey)

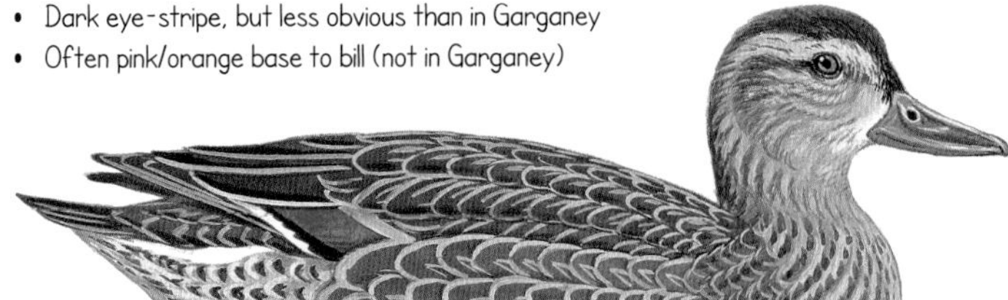

Garganey

- Longer body and tail than Teal's
- Stripy head

Teal

- White streak under tail
- 'Double-scaled' flank-feathers
- Green speculum often visible

Garganey

- Slightly longer neck than Teal's
- Overall paler wings than Teal's
- Relatively broad white trailing edge
- Thin white greater wing-covert bar

Teal

- Dark outer wing
- Narrow white trailing edge to wing
- Inner white wing-bar broader than trailing edge (reverse in female Garganey)
- Iridescent green speculum

FEMALE *AYTHYA* DIVING DUCKS

INTRODUCTION Tufted Duck (*Aythya fuligula*) [L 43cm] and Common Pochard (*Aythya ferina*) [L 46cm] common and widespread on freshwater ponds and lakes. Greater Scaup (*Aythya marila*) [L 46cm] breeds in Scandinavia and Iceland on freshwater pools and lakes within forest-tundra and tundra zones; in winter on sheltered, shallow coasts. Ferruginous Duck (*Aythya nyroca*) [L 41cm] rare on freshwater lakes with fringing vegetation.

Female diving duck silhouettes

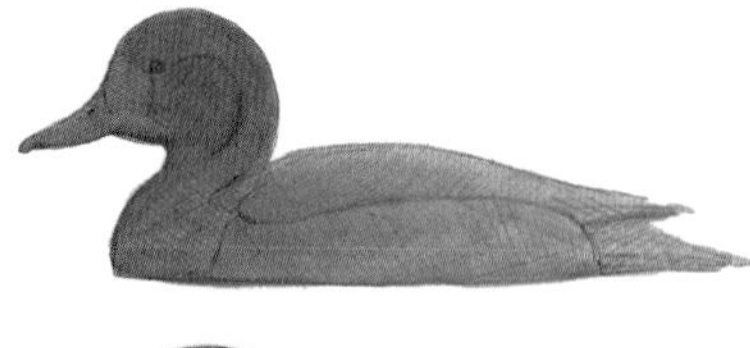

Scaup

- Slightly larger than Tufted; roughly equal to Common Pochard
- Evenly rounded, rather large head without tuft

Tufted Duck

- Tuft or hint of tuft
- High forehead
- Steep nape below flattish crown
- Quite broad but short bill

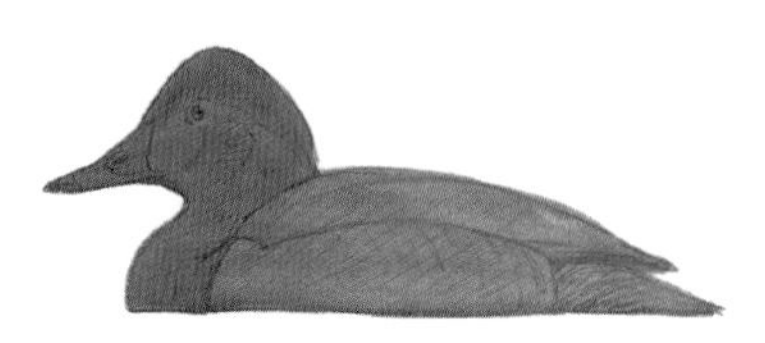

Pochard

- Large bill with upwards curve to culmen that continues up to crown, creating even, 'ski-jump' slope
- Notable peak at crown
- Gentle slope down nape, giving triangular shape to head
- Larger and longer bodied than Tufted

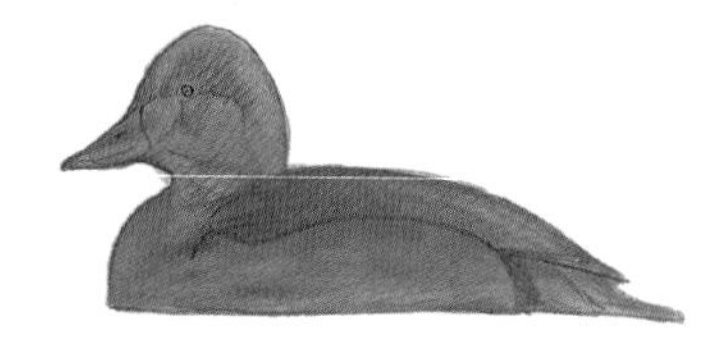

Ferruginous Duck

- Tufted-sized but notably short bodied
- Longish, broad-based bill
- Peaked crown similar to Common Pochard's, but steeper forehead and nape (squashed pochard)

Common Goldeneye

- Small, conical bill
- Very steep forecrown
- Outsize head (looks too big for neck); even slope down nape
- Low in water with sloping back

Sleeping female *Aythya* diving ducks

Scaup

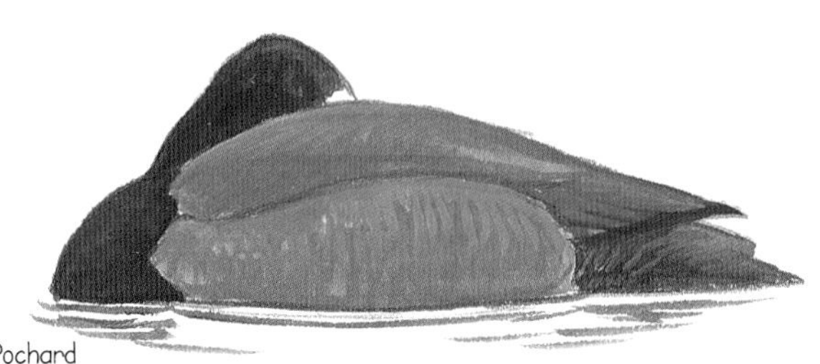

- Bulkier than similar Tufted
- Back slopes down evenly to water (c.f. Tufted)
- Breast often protrudes more than in Tufted
- Greyer above than Tufted
- Darker rear end than in Common Pochard

Pochard

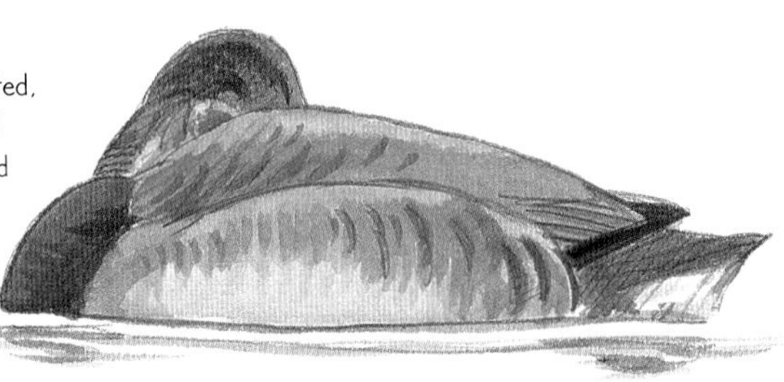

- Paler head than that of Scaup or Tufted, with hints of 'powder patches' on head
- Rustier breast than in Scaup or Tufted
- High crown and odd head shape obvious even in sleeping birds
- Upperparts and underparts of similar colour (c.f. Tufted)

Tufted Duck

- Dark upperparts contrast with well-demarcated, pale dusky flanks
- Some (not all) individuals have patch of white on undertail-coverts - not usually as prominent as in Ferruginous, and generally sullied by buff marks

Ferruginous Duck

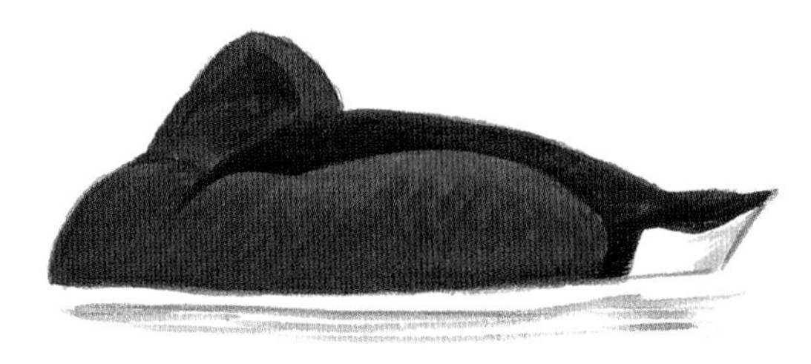

- Rich chestnut colour
- Peaked crown with 'steep sides'
- Brilliant white, large and well-marked undertail patch

Female Scaup and Tufted Duck

Tufted Duck (female)

- Small tuft on back of neck a clincher (variable - sometimes just bump)
- Unsullied dark brown back (c.f. Scaup)
- Well-defined flank panel, paler than rest of plumage

Scaup (female)

- Flanks greyer than in Tufted, with diffuse scaling
- Back brownish-grey with fine pencil-line vermiculations (fine barring)
- Just a hint here of pale 'ear-muffs' that appear quite clearly in spring and summer

Tufted Duck (typical female)

- Fairly typical Tufted female with reduced white on head

Scaup (female)

- Broad bill
- Bill mainly grey, just with black nail
- Copious white covering whole bill-base, including well up forehead

Tufted Duck (female, variation)

- Some females have white at bill-base. Few have as much as this, well wrapped around whole bill - but check bill pattern and head shape.
- Relatively extensive black at bill tip

Scaup (juvenile, summer)

- Similar to female Scaup, but less white at bill-base
- Often has darker bill with less obvious black nail

Scaup (late 1st-winter male)

- Looks like female Scaup with black head and no white at bill-base; also rather similar to Tufted
- Back often looks messy, with mixture of brown-and-white scaling as dark brown giving way to neat grey back of adult male
- Grey flanks with some brown scaling

Tufted Duck/Pochard hybrid (male)

- Tuft on head (as Tufted)
- Reddish hue to head (as Common Pochard)
- Grey back and white sides (as Scaup)

Tufted Duck (male)

- Purple sheen to head
- Large tuft
- Gleaming white, well-defined flank panel

Scaup (female)

- White at bill-base broad and even - looks like nose-band
- Black at tip restricted to nail; oval in profile
- Broad beamed and bulky compared with Tufted

Tufted Duck (female)

- Individual with white at bill-base; white tends to be narrow and uneven
- Whole end of bill black
- Smaller head than Scaup's

Tufted Duck (female)

- Common sight - look out for broad white wing-bar and dashing flight with very fast wingbeats (Scaup less 'whirring')
- More compact than female Scaup

Scaup (female)

- Striking wing-bar, as in Tufted
- Broader wings than in Tufted, giving distinctly bulkier appearance
- Looks paler than Tufted
- White face often stands out in flight

Ageing Ferruginous Duck

Ferruginous Duck (adult female)

- Overall rich burnt-chestnut colour; little contrast between upperparts and underparts
- Bill fairly long, dark grey, with darker nail and, usually, whitish band before tip (but can vary)
- Clean white undertail-coverts, well defined with thin black line
- Flank feathers often ruffle higher up sides than in Tufted
- Dark eye (c.f. Tufted)

Ferruginous Duck (juvenile)

- Darker than female Ferruginous
- Flanks scalier than adult female's
- Whitish throat
- White on undertail not usually as bright as in adult female

SCOTERS

INTRODUCTION Common Scoter (*Melanitta nigra*) [L 49cm] breeds on northern tundra freshwater lakes and rivers, mainly in Scandinavia; widespread inshore coastal areas in winter. Velvet Scoter (*Melanitta fusca*) [L 54cm] breeds on freshwater lakes and pools in wooded areas of taiga; as Common in winter. Surf Scoter (*Melanitta perspicillata*) [L 51cm] rare visitor from North America; typically a stray with scoter flocks.

Scoters are diving ducks with predominantly black plumage. They are often spotted in large rafts offshore on sheltered coasts during winter.

Head profiles

Velvet Scoter

- More or less uninterrupted slope from crown to bill
- Bill large and wedge shaped
- Head looks bigger than Common's
- Note thickness of neck, especially lower neck

Common Scoter

- Sloped forehead makes obvious angle with bill, especially slight bulge of male's culmen
- Rounded nape
- Looks pert and small headed

Female scoters

Velvet Scoter

- Some white usually visible on wing, even when swimming or loafing
- Usually pale patch at bill-base
- Usually clear patch on rear of ear-coverts (may be larger than patch at bill-base)

Common Scoter

- Well-defined, dusky greyish-brown cheeks and neck sides

Surf Scoter

- Clear white patch at bill-base, much more vertical than any patch on adult female or juvenile Velvet
- White streak behind eye (recalls juvenile Velvet)
- Some have white patch on nape
- Heavy, swollen bill with broad base

Scoters in flight

Common Scoter (male)

- Outer wing has silvery sheen
- Small yellow patch on upper mandible (culmen)
- Knob at top of culmen
- Dark legs (c.f. Velvet, Surf)

Common Scoter (female)

- Like sooty-brown version of male, except for distinctive head and neck pattern
- Head and neck pale grey-brown; head pattern completely different from any other scoter's

Common Scoter (1st-winter male)

- Overall, combines black body plumage of breeding male with brown juvenile wing feathers
- Whitish belly typical of subadult scoters
- Lacks knob on bill, but has variable yellow

Velvet Scoter (male)

- Brilliant white secondaries form neatly squared-off wing-panel
- White comma shape near eye
- Relatively large patch of yellow on lower mandible (not upper one, as in Common)

Velvet Scoter (female)

- White wing-patch every bit as obvious as male's
- Body colouring dusky brown

Velvet Scoter (1st-winter male)

- Blacker than female
- Traces of white belly

Surf Scoter (male)

- All-black wings lack silvery sheen (c.f. Common)
- Head looks relatively enormous (almost like puffin's or other auk's)
- Large white patch on back of neck

Surf Scoter (female)

- Unmarked wings
- Slightly darker brown than Common (but similar size)
- Enormous swollen grey bill, broad at base; often first thing noticed is heavy bill and head

Surf Scoter (1st-winter male)

- White belly distinguishes it from adults
- Cheeks and neck paler than female Surf or Velvet's (closer to female Common's)

General scoter ID tips

Velvet Scoter (adult female, stretching)

- Attitude taken by Velvet (and Surf) when wing flapping (Common Scoter puts head down)
- White wing-panel

Velvet Scoter (1st-winter male)

- (September-March)
- Hint of paler feathering at bill-base and ear-coverts
- Dark eye (c.f. adult)
- Yellow bill (as adult).

Surf Scoter (adult male)

- Huge white nape-patch
- White eye
- Bill tends to look orange at distance

Surf Scoter (female)

- Some have white patch on nape
- White patch near bill more elongated than in Velvet
- Broad-based bill with fairly even slope to culmen (similar to Common Eider)
- Darker head than female Common Scoter, without darker cap

Common Scoter (female)

- Largely unmarked, dark sooty-brown plumage, although slightly darker above
- Contrastingly pale cheek and neck side, sharply demarcated
- Clearly defined black cap
- Relatively small, greyish bill

EIDER

INTRODUCTION Common Eider (*Somateria mollissima*) [L 64cm] widespread and numerous along coasts throughout the region, and often abundant in the north.

Males take 3–4 years to mature, displaying a variety of intermediate plumages as they do so.

Eider in flight

Ageing Eider

JUVENILE

July-September

- Very dark
- Narrow, dense bars
- Pale supercilium

IMMATURE, AUTUMN

September-November

- Even darker than juvenile
- Supercilium not as obvious, if at all

MALE, ECLIPSE

July-October, much individual variation in timing and appearance

- White tertials (moulted August), otherwise dark

EARLY 1ST-WINTER MALE

October-December

- White begins to develop on breast first

1ST-WINTER MALE

December-March

- Whitens on flanks, back and head

1ST-SUMMER MALE

April-June

- Black belly develops
- Head still dark

LATE 1ST-SUMMER MALE

June-July, then similar in following winter

- Head gradually whitens
- Bill shows signs of green
- Back largely white

SUBADULT MALE, SPRING, 2 YRS OLD

roughly March-July

- Similar to adult, but lacks curved white tertials
- White plumage has dark blotches

ADULT MALE

All year except July-October, when it is in eclipse, moulting plumage

- White above, dark below - reverse of many seabirds
- Green nape (unique)
- White tertials and thighs
- Black crown

ADULT FEMALE, TYPICAL

- Almost Mallard coloured, but with huge, wedge-shaped pale bill
- Dark speculum with pale edges (distinguishes it from juvenile)
- Close barring (female Mallard more streaked than barred)

ADULT FEMALE, RUFOUS MORPH

- Huge, triangular head, with bill fitting neatly into skull in straight line
- 'Lobe' of bill reaches up forehead and feathering reached as far as nostril (unlike vagrant female King Eider)

FEMALE SAWBILLS

INTRODUCTION The 'sawbills' are a small group of fish-eating ducks that have serrated bill edges to help them hold on to prey. Female sawbills are often called 'red-heads'. Sawbills often dive for long periods and may suddenly reappear well away from where they submerged. Goosander (*Mergus merganser*) [L 63cm] breeds Scandinavia and northern Britain, beside forested lakes and along rivers. In winter widespread on large freshwater lakes and rivers. Red-breasted Merganser (*Mergus serrator*) [L 55cm] has similar distribution; nests by fresh and salt water. In winter throughout the region on inshore waters and estuaries. Smew (*Mergellus albellus*) [L 41cm] scarce breeder in northern Scandinavia by taiga lakes; in winter localized on freshwater lakes.

GOOSANDER

- Big, long, powerful and impressive duck - an 'aircraft carrier' among ducks
- Head looks large
- Sharp division between dark head and whitish neck (probably best distinction from Red-breasted Merganser)
- Prominent white throat

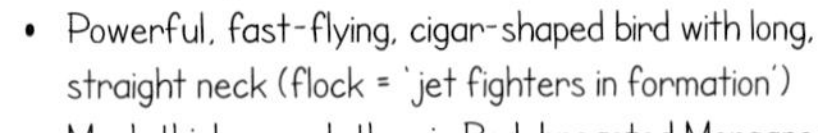

- Powerful, fast-flying, cigar-shaped bird with long, straight neck (flock = 'jet fighters in formation')
- Much thicker neck than in Red-breasted Merganser
- Often hint of pink on breast and belly (not in Red-breasted Merganser)

- Crest tends to look neater and less spiky than Red-breasted Merganser's - more like mane or bob (but can look spiky when raised)

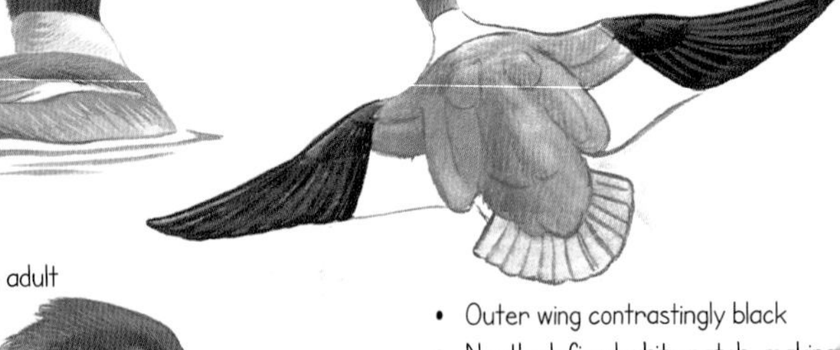

JUVENILE

- Not as richly brown on head as adult
- Bill paler than adult's
- White loral line often shows (resembles that of Red-breasted Merganser)
- Paler eye than adult's

- Outer wing contrastingly black
- Neatly defined white patch, making a 'square' on inner wing; some birds may show hint of black dividing line (c.f. Red-breased Merganser)

Smew

- Small and compact
- Bill grey and very different in shape from bills of other species: shorter, and lacking 'tooth' at tip
- White wing-patch sometimes shows in mid-body (note similarity to female Goldeneye)
- Chestnut-brown crown and nape
- Brilliant white cheeks contrast with crown

- White ovals near base of wing
- Narrow white trailing edge formed by secondary tips

- Upperparts very grey
- White panels in mid-wing, not trailing as in the other species
- Wings very black and white

Red-breasted Merganser

- Darker body than Goosander's
- Poorly differentiated whitish throat and front of neck
- Diffuse but noticeable pale lores
- Bill often shows slight upcurve

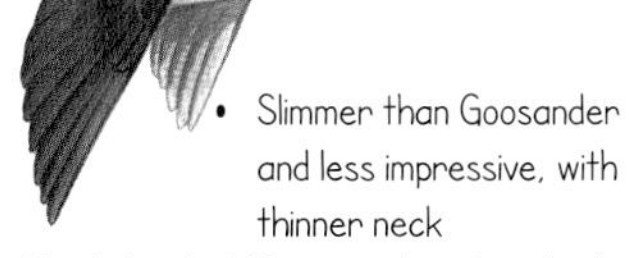

- Slimmer than Goosander and less impressive, with thinner neck
- Head clearly different colour from body, but contrast not sharp
- Somewhat paler, subtler head colour than in Goosander

- Crest very different from Goosander's - like Goosander with unprofessional haircut. Spiky at nape and rear of crown, and sometimes very prominent, but at times hardly noticeable at all
- Bill much thinner than Goosander's, especially at base, and hook less obvious

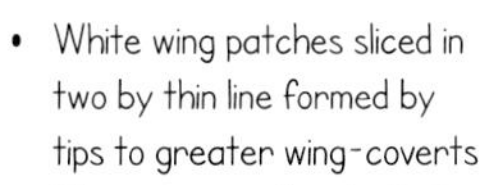

- Crest may look alarmingly spiky and punk-like, as if bird has washed its hair and let the wind dry it
- Plastic-looking, long bill may give bird appearance of toy duck

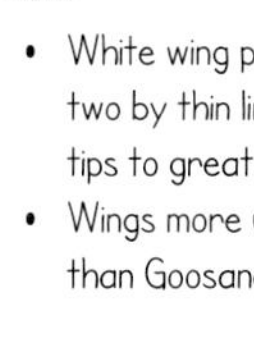

- White wing patches sliced in two by thin line formed by tips to greater wing-coverts
- Wings more uniformly dark than Goosander's

WINTER DIVERS

INTRODUCTION Red-throated Diver (*Gavia stellata*) [L 62cm] breeds Scandinavia, Iceland and northern Britain on freshwater pools, tundra lakes and marine inlets. Winters widely on inshore waters. Black-throated Diver (*Gavia arctica*) [L 69cm] on larger and deeper lakes than Red-throated, and not in marine inlets. Winters as Red-throated. Great Northern Diver (*Gavia immer*) [L 80cm] breeds Iceland on large freshwater lakes. Winters Atlantic and North Sea coasts.

To dive, divers usually slip down head-first with barely a ripple, while Cormorants and small grebes leap. In flight, grebes look weak and desperate to land. Divers are powerful fliers and often rise high above the water, which grebes do not do.

Differences between grebes, divers and cormorants

Great Crested Grebe

- Abrupt end to rear
- Never holds heads uptilted when swimming (neither does any grebe)

Red-throated Diver

- Longer body than grebe's, and at least suggestion of tail
- Thicker neck than grebe's
- Front parts look more solid than grebe's
- Swims lower in water than grebe

Cormorant

- Holds head up when swimming
- Blunt-tipped bill
- Long stern and tail
- Angled back of head
- Jumps into dive (unlike diver)

Divers at distance

Great Northern Diver

- Dagger-shaped, sharp-tipped bill
- Strong contrast between black crown and nape, and white throat and breast
- High crown and 'bump' on head (c.f. other divers, Cormorant)

Black-throated Diver

- Smart and contrasting
- Thick bill, like dagger; holds head horizontal
- White patch on flank

Red-throated Diver

- Head held up at slight angle when swimming (as Cormorant, c.f. other divers)
- Vertical drop down to chest
- Very white on head and neck
- Paler than other divers
- Thinner bill than other divers

Divers at close range

Red-throated Diver

- Flatter forehead than in other divers
- Whitish spots/scales on back
- Stripes on hindneck may be visible (c.f. other divers)

Black-throated Diver

- Very clean contrast on neck (see Great Northern), 50/50 dark/white
- Greyish colour on back of neck clearly paler than upperparts of back or wings (reverse of Great Northern)
- Curved neck
- 'Rising moon' white mark on flank
- Slight comma on cheek edge

Great Northern Diver

- Pale area around eye (c.f. Black-throated)
- Ear-coverts diffusely mottled
- Dark half-collar on lower neck
- Plumage can look a little messy

Diver ID tips (both images Great Northern Diver)

- When splayed, feet can look enormous
- Divers often raise themselves out of water to flap wings. Cormorants do not do this; grebes only rarely

Heads of winter divers

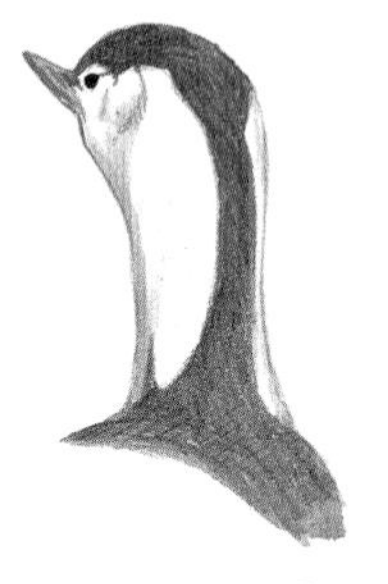

Black-throated Diver

- Dark grey coloration covers neck

Great Northern Diver

- White 'neck nick' can be conspicuous
- Bump on forehead (Black-throated sometimes shows this)
- White corona around eye
- Neck decidedly thick

Red-throated Diver

- Very thin dark nape ('short back and sides')
- Eye clearly visible

Divers in flight

Great Crested Grebe (for comparison)

- Large white wing-bars and panels (larger grebes have white on upperwings, divers do not)
- Slimmer neck than diver
- Fast wingbeats and weak impression; divers slower and powerful

Red-throated Diver

DISTANT

- Rather small head just a blob, and thin bill adds to slender profile
- Chin sometimes seems to bulge

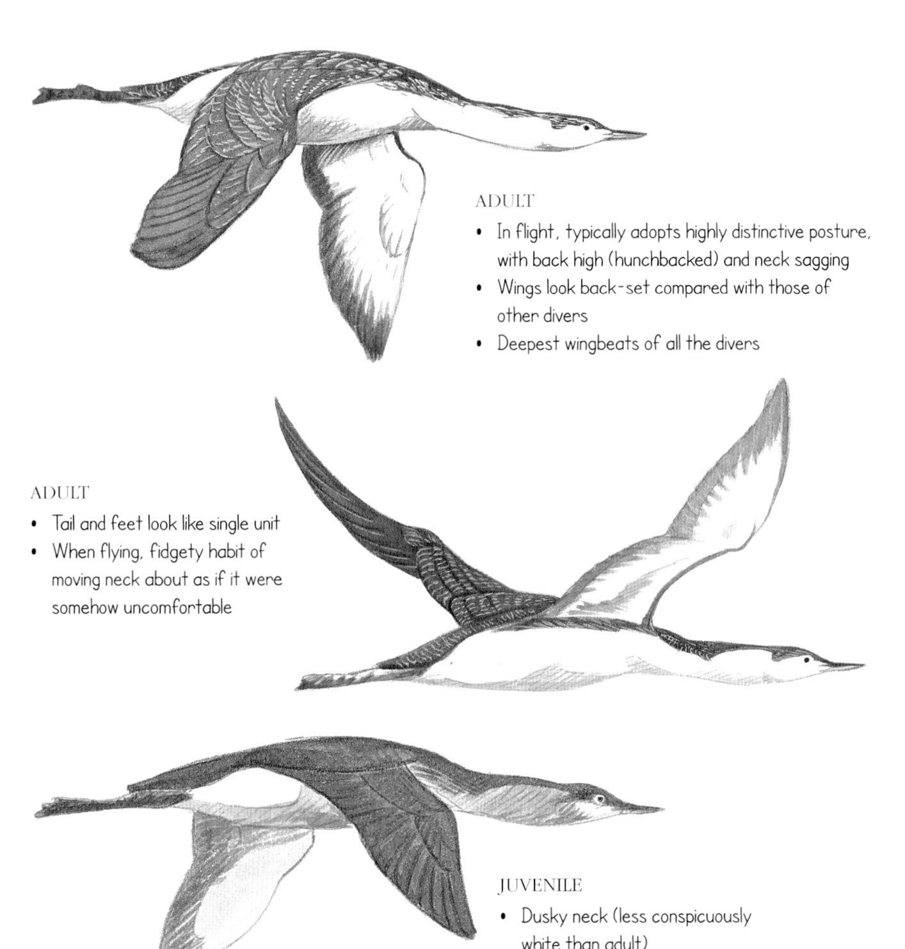

ADULT

- In flight, typically adopts highly distinctive posture, with back high (hunchbacked) and neck sagging
- Wings look back-set compared with those of other divers
- Deepest wingbeats of all the divers

ADULT

- Tail and feet look like single unit
- When flying, fidgety habit of moving neck about as if it were somehow uncomfortable

JUVENILE

- Dusky neck (less conspicuously white than adult)

Black-throated Diver

DISTANT

- In contrast to Red-throated, neck held up straight in flight (rather than sagging)
- Tends to look more elongated than Red-throated (more projecting feet)

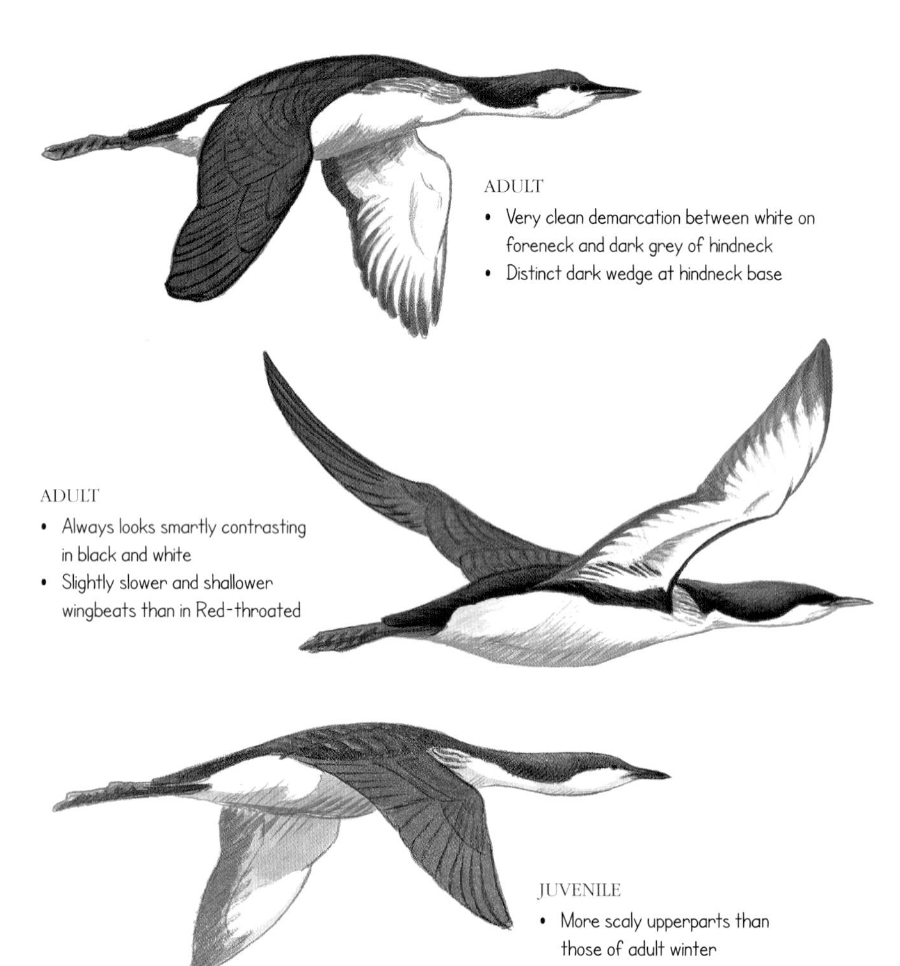

ADULT

- Very clean demarcation between white on foreneck and dark grey of hindneck
- Distinct dark wedge at hindneck base

ADULT

- Always looks smartly contrasting in black and white
- Slightly slower and shallower wingbeats than in Red-throated

JUVENILE

- More scaly upperparts than those of adult winter

Great Northern Diver

DISTANT

- Feet often (not always) look big, and trail far behind
- Noticeably thick neck
- Belly plumper than in Black- or Red-throated
- In flight, tends to keep head up (see Red-throated), and thus has straight profile

ADULT

- Slower wingbeats than those of other divers, with slightly more relaxed impression; flight has been described as goose-like, and does not fit other diver species

ADULT

- Large diver that can recall Cormorant in flight (others tend not to)
- Inner part of wing (arm) broader than that of Black- or Red-throated

JUVENILE

- Upperparts more scaled than in adult
- Paler brown neck

FULMAR AND SHEARWATERS

INTRODUCTION Northern Fulmar (*Fulmarus glacialis*) [L 46cm, WS 110cm] breeds commonly on cliffs and islands around the British Isles, Norway, France and Iceland; when not breeding wanders North Sea and Atlantic. Manx Shearwater (*Puffinus puffinus*) [L 32cm, WS 78cm] breeds on inaccessible and rat-free islands off British Isles, France and Iceland, March–September; winters Southern Oceans. Cory's Shearwater (*Calonectris borealis*) [L 48cm, WS 120cm] breeds southern Europe in the Atlantic and Mediterranean, but disperses north July–September and is regular off Ireland and south-west Britain; rare otherwise. Similarly, Balearic Shearwater (*Puffinus mauretanicus*) [L 37cm, WS 85cm] breeds in the western Mediterranean and disperses north in late summer, mainly to the English Channel and Atlantic coast of France. Sooty Shearwater (*Puffinus griseus*) [L 44cm, WS 102cm] breeds in the South Atlantic but roams north in our summer, turning up in the North Atlantic and even North Sea regularly in July–October. Great Shearwater (*Puffinus gravis*) [L 46cm, WS 112cm] as Sooty, but rare in the North Sea.

Cory's Shearwater

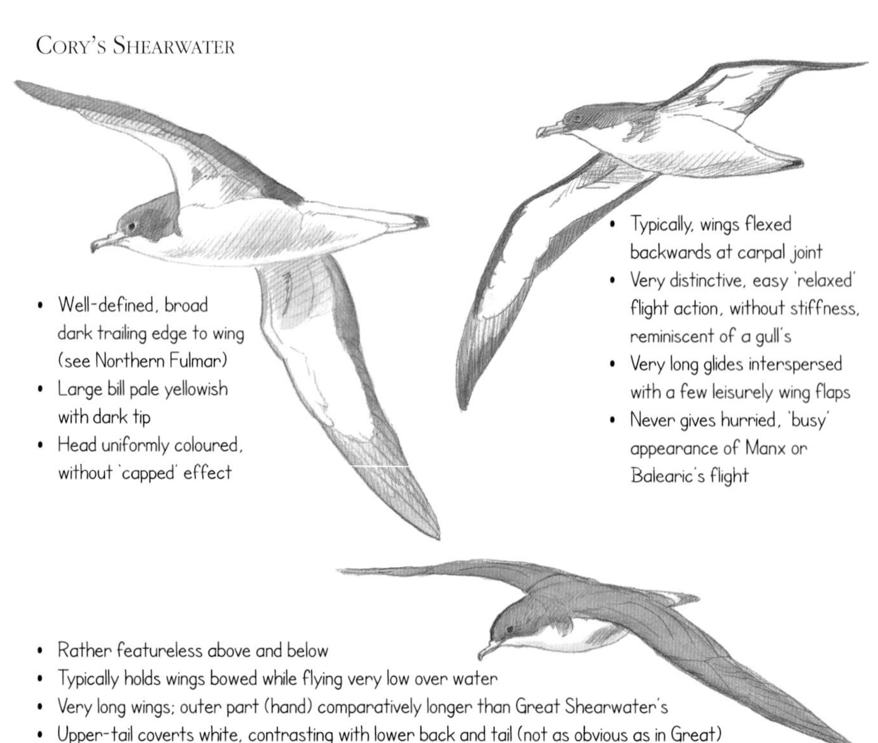

Northern Fulmar

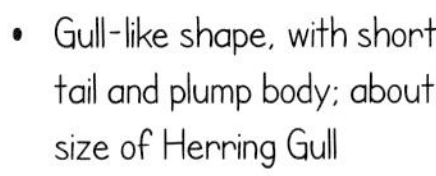

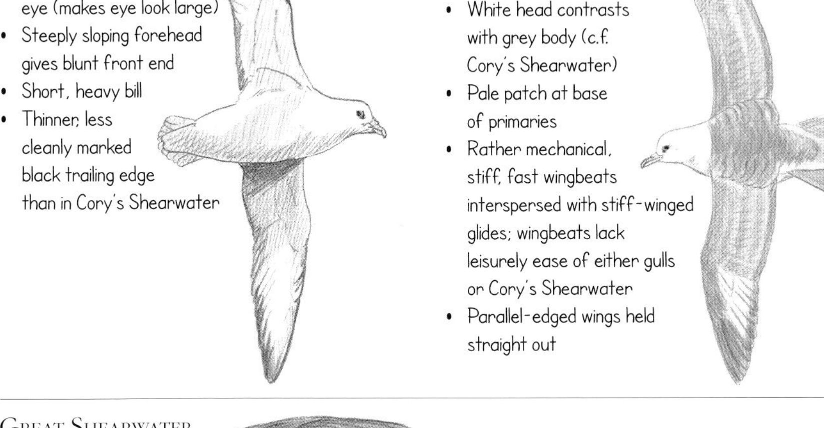

- Dark patch in front of eye (makes eye look large)
- Steeply sloping forehead gives blunt front end
- Short, heavy bill
- Thinner, less cleanly marked black trailing edge than in Cory's Shearwater

- Gull-like shape, with short tail and plump body; about size of Herring Gull
- White head contrasts with grey body (c.f. Cory's Shearwater)
- Pale patch at base of primaries
- Rather mechanical, stiff, fast wingbeats interspersed with stiff-winged glides; wingbeats lack leisurely ease of either gulls or Cory's Shearwater
- Parallel-edged wings held straight out

Great Shearwater

- Slightly darker upperparts than Cory's
- Not much smaller than Cory's, but looks slimmer and neater
- Slightly darker upperparts than Cory's
- Often shows greyish cast to inner wing (arm)
- Upper-tail coverts distinctively and contrastingly white, usually more clearly so than in Cory's
- Wing-tips more sharply pointed than Cory's

- Axillaries ('arm-pits') smudged with complex markings (c.f. Cory's)
- Dark patch on white belly
- Bill entirely dark (c.f. Cory's)
- Undertail-coverts dark (c.f. Cory's)

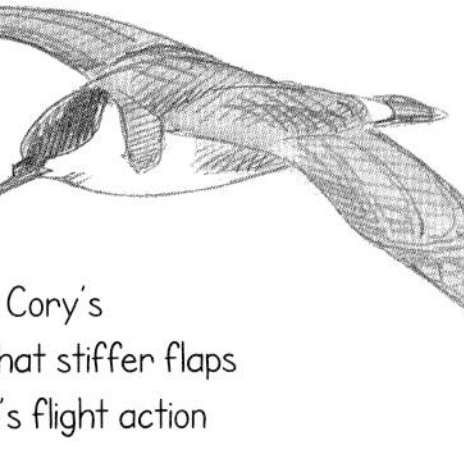

- Distinctive large dark smudge on shoulder
- White collar visible at long range
- Distinctive black cap
- Wings not held distinctively bowed like those of Cory's
- Flight quite different from Cory's, with somewhat stiffer flaps in between glides; not as stiff-winged as Manx's flight action

Manx Shearwater

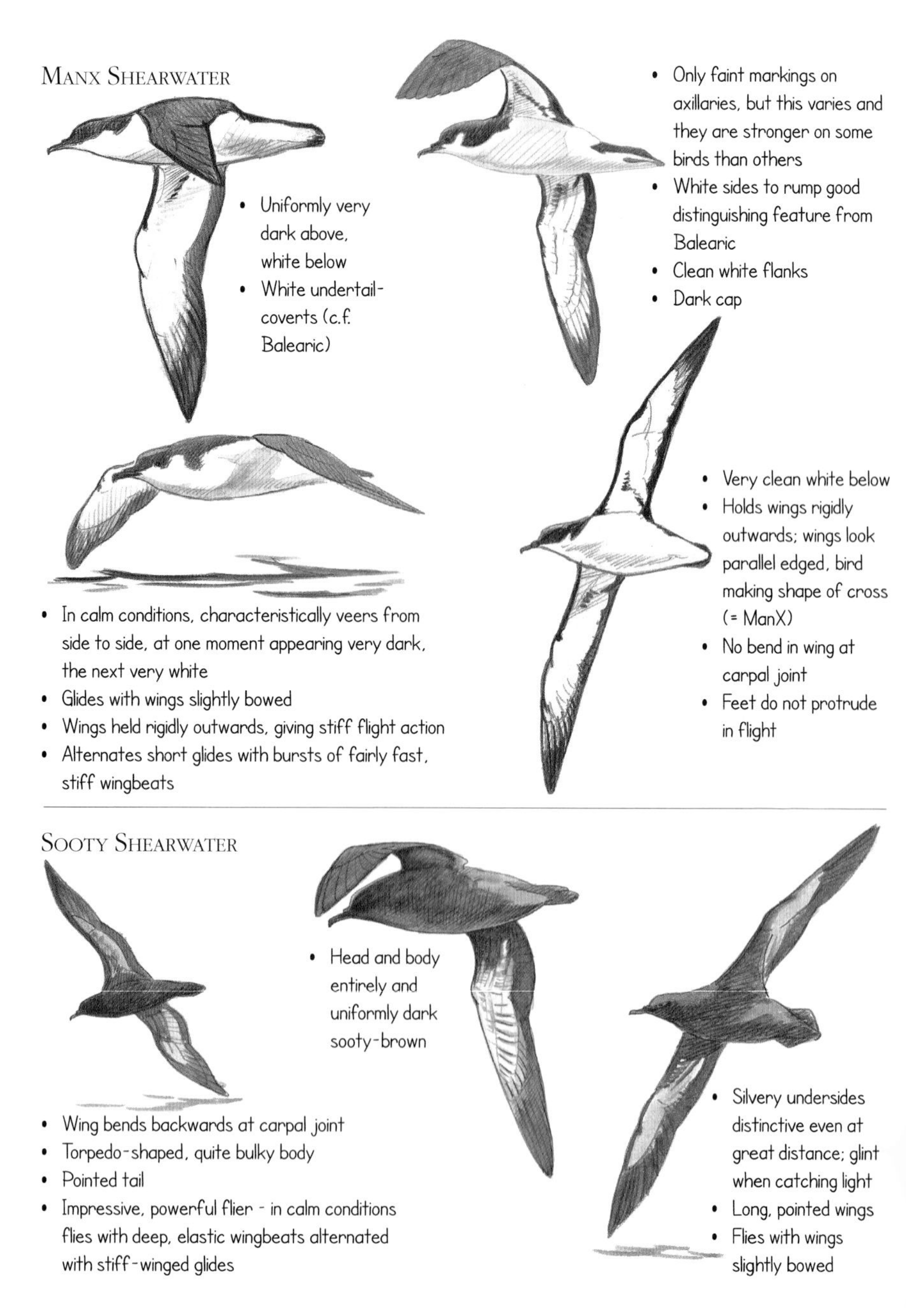

- Uniformly very dark above, white below
- White undertail-coverts (c.f. Balearic)

- Only faint markings on axillaries, but this varies and they are stronger on some birds than others
- White sides to rump good distinguishing feature from Balearic
- Clean white flanks
- Dark cap

- In calm conditions, characteristically veers from side to side, at one moment appearing very dark, the next very white
- Glides with wings slightly bowed
- Wings held rigidly outwards, giving stiff flight action
- Alternates short glides with bursts of fairly fast, stiff wingbeats

- Very clean white below
- Holds wings rigidly outwards; wings look parallel edged, bird making shape of cross (= ManX)
- No bend in wing at carpal joint
- Feet do not protrude in flight

Sooty Shearwater

- Head and body entirely and uniformly dark sooty-brown

- Wing bends backwards at carpal joint
- Torpedo-shaped, quite bulky body
- Pointed tail
- Impressive, powerful flier - in calm conditions flies with deep, elastic wingbeats alternated with stiff-winged glides

- Silvery undersides distinctive even at great distance; glint when catching light
- Long, pointed wings
- Flies with wings slightly bowed

Balearic Shearwater

- Distinctly pale brown upperparts, not as dark as in Manx
- In flight holds wings out rigidly and progresses with stiff wingbeats (good distinction from Sooty)

- Much plumper than Manx - noticeable even when bird is distant

- Much duskier below than Manx often looks a bit dirty. Some individuals much darker than this and may resemble Sooty
- Always has pale patch on belly, distinguishing even darkest birds from Sooty; also lacks glinting wing underside of Sooty
- Quite strong pattern on underwings, especially compared with Manx
- Often has dark undertail-coverts, an easy distinguishing feature from Manx

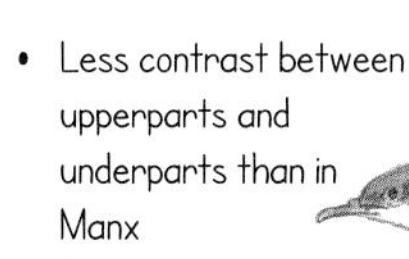

- Less contrast between upperparts and underparts than in Manx
- Duskier throat than in Manx, without same contrast to cap
- Projecting feet

Shearwater silhouettes

Manx Shearwater

- Wings held straight out, without much bend at carpal joint (c.f. Sooty)
- Slim and neat
- Shorter wings than Sooty's

Balearic Shearwater

- Distinctively pot bellied ('Balearic barrel')
- Shorter tail than in Manx or Sooty
- Legs project beyond tail

Sooty Shearwater

- Wings backswept at carpal joint, especially in strong wind
- Body larger and much bulkier than Manx's

GANNET

INTRODUCTION Northern Gannet (*Morus bassanus*) [L 93cm, WS 172cm] wide-ranging seabird commonly seen not far offshore. Breeds in scattered colonies in the North Sea and Atlantic. Flies with slow wingbeats interspersed with long glides. Wings are usually held straight out, but look swept back in strong winds.

Ageing Gannet

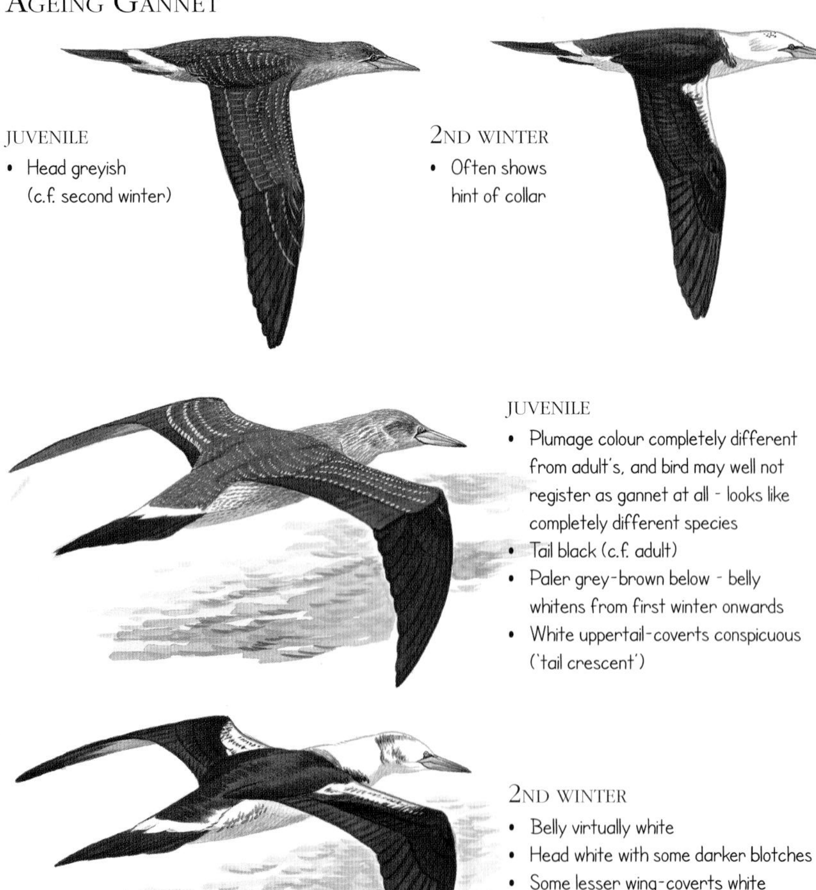

JUVENILE

- Head greyish (c.f. second winter)

2ND WINTER

- Often shows hint of collar

JUVENILE

- Plumage colour completely different from adult's, and bird may well not register as gannet at all - looks like completely different species
- Tail black (c.f. adult)
- Paler grey-brown below - belly whitens from first winter onwards
- White uppertail-coverts conspicuous ('tail crescent')

2ND WINTER

- Belly virtually white
- Head white with some darker blotches
- Some lesser wing-coverts white

2ND SUMMER

- Belly now entirely white
- Adult pattern on head and neck
- Tail still black

3RD WINTER

- For the first time gains adult's butterscotch-yellow coloration on head; colour brightest in spring, as here
- Wings similar to adult pattern, but blotchy

4TH WINTER

- Head same yellowish colour as in adult - colour fades in winter months
- Central tail feathers black (c.f. adult)
- Secondary bar along trailing edge of wing can have 'piano-keyboard' pattern

ADULT

- Tail all white (c.f. immatures)
- All gleaming white but for black wing-tips and black marks around eye, plus yellowish head

CORMORANT & SHAG

INTRODUCTION Great Cormorant (*Phalacrocorax carbo*) [L 85cm] widespread throughout on sea coasts and inland wetlands. European Shag (*Phalacrocorax aristotelis*) [L 73cm] exclusively a seabird; mainly rocky coasts of Britain, Ireland, France, Iceland and Norway.

Note that adult Cormorants vary greatly in size and some females are not much bigger than Shags; some Cormorants have smaller and slimmer bills than average.

Ageing Shag

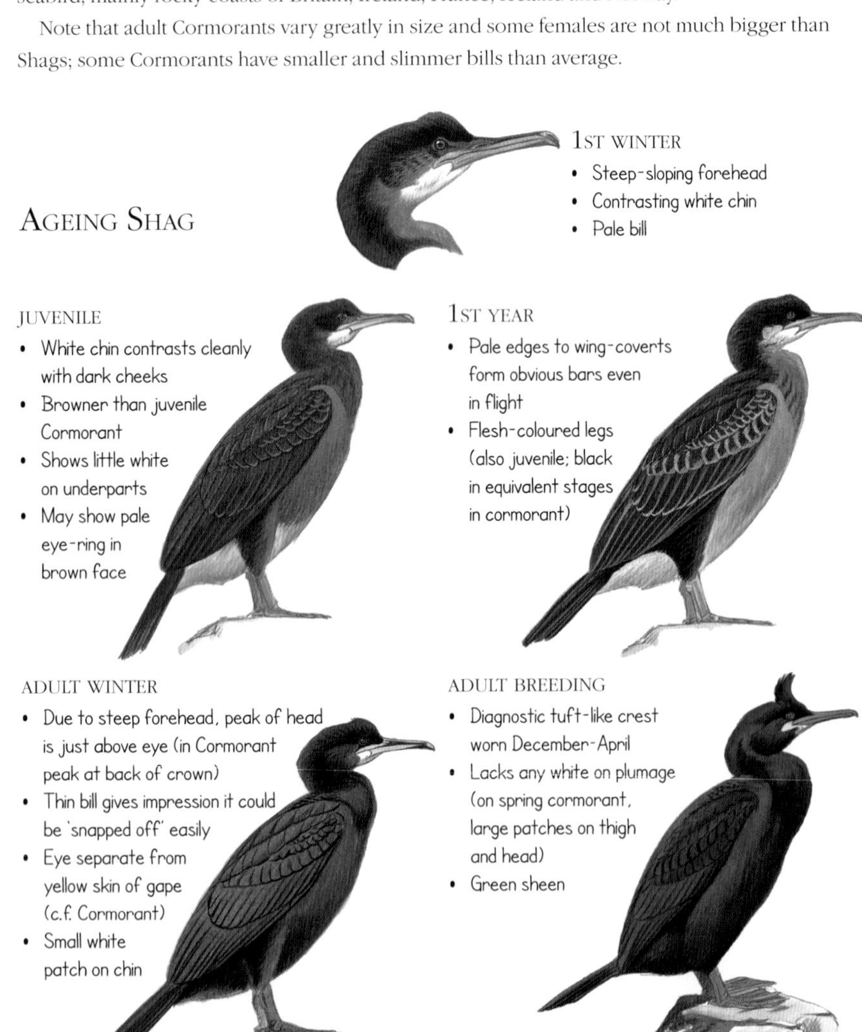

Flight silhouettes

Shag

- Neck held straighter than Cormorant's, without kink
- More pot bellied than Cormorant
- Wingbeats faster and looser; does not glide (c.f. Cormorant)
- Slightly more rounded wings than Cormorant's

Cormorant

- Similar to goose in flight, but has longer tail, is silent and often intersperses flaps with glides
- Quite quick but shallow wingbeats; sometimes soars high in sky
- Thicker head and neck than Shag's

Ageing Cormorant

Juvenile

- Often (but not always) shows substantial areas of white on underparts - much variation, with some individuals almost black
- Diffuse white cheeks
- Grey bill

1st winter

- Often shows bronze tint to upperparts (c.f. Shag)

1st winter–1st summer

- 'Wrap-around' bare yellow (sometimes orange) skin at base of bill
- Scaling already obvious on upperparts
- Dark belly

Adult winter

- Thick bill 'sunk deeply' into head
- Thick, wedge-shaped head with flat crown
- Eye within bare skin of gape
- Large white patch on sides of face

WINTER GREBES

INTRODUCTION Great Crested Grebe (*Podiceps cristatus*) [L 48cm] widespread on freshwater lakes and slow-flowing rivers; winters in similar habitats, and sheltered coasts. Red-necked Grebe (*Podiceps grisegena*) [L 43cm] scarce breeder in the western Baltic region, including Denmark, northern Germany and southern Sweden, on well-vegetated lakes and rivers; in winter more widespread, also on reservoirs and sheltered coasts. Little Grebe (*Tachybaptus ruficollis*) [L 26cm] widespread on well-vegetated marshes, rivers, lakes and ponds; in winter more open habitats, including estuaries. Slavonian Grebe (*Podiceps auritus*) [L 34cm] breeds Scotland, Iceland and Scandinavia on freshwater lakes; more widespread in winter and tends towards sheltered coasts in particular. Black-necked Grebe (*Podiceps nigricollis*) [L 31cm] breeds further south than Slavonian, but very locally on marshes and freshwater lakes with much emergent vegetation. More widespread in winter, also on reservoirs, estuaries and sheltered coasts.

In flight, all the grebes progress with very fast, shallow wingbeats, the smaller species giving a similar impression to auks, only weaker and more reluctant; flying always seems hard work for these birds. Slavonian and Black-necked Grebes in flight look 'like plucked chickens on skewers'.

Small grebe silhouettes

Little Grebe

- Prominent, fluffy rear end - small, floating ball of feathers
- High back when relaxed; looks much slimmer when diving actively
- Short, weak bill with obvious gape

Black-necked Grebe

- High, fluffy-backed stern (c.f. Slavonian)
- Rounded head with steep forehead
- Thin, upwardly tilted, sharp-pointed bill

Slavonian Grebe

- Small but neatly proportioned
- Less fluffy ended than other small grebes
- Straight neck
- Flatter crown than in other small grebes, with gently sloping forehead
- Straight, dagger-shaped bill

Heads

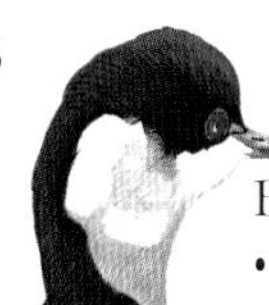

Black-necked Grebe

- Dusky cheeks
- Black below eye

Slavonian Grebe

- Gleaming clean white cheeks
- Thinner black line down nape

Small grebes from behind

Little Grebe

- Browner than other small grebes
- Fairly weak contrast between cap and cheeks
- Dark eye

Black-necked Grebe

- Higher rear end than in Slavonian
- 'Riding-hat' pattern made by black on cap and cheeks

Slavonian Grebe

- Relatively flat crown
- Cleaner and neater contrast between black crown and nape, and whiter cheeks and neck than in Black-necked
- Often visible pale patch on lores

Small grebes from side

Black-necked Grebe

Black on crown extends down cheeks
Tongue of white loops back behind ear-coverts

Slavonian Grebe

- Black/white contrast in relatively straight line across cheeks
- White bill tip

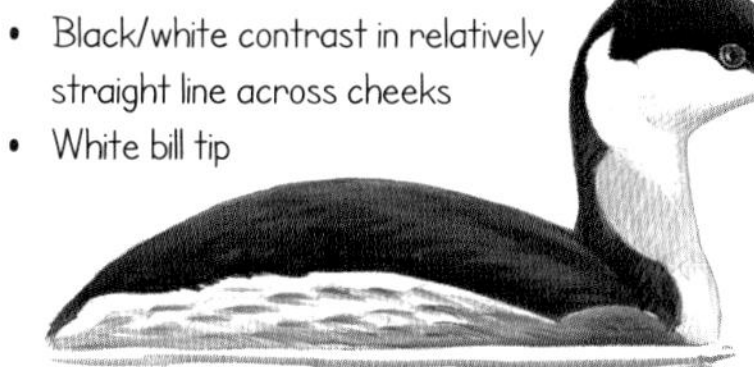

Little Grebe

- Buff-brown underparts (not white)
- Thick, relatively blunt yellowish bill
- Obvious gape

Grebes in flight

Great Crested Grebe (landing)

- Grebes always 'crash-land' on water, belly-first

Little Grebe

- Often appears barely capable of flying - tends to skitter with whirring feet over water's surface, wings flapping madly
- No clearly visible white panels on wing - narrow trailing edge at best (c.f. Black-necked, Slavonian)
- Some younger birds have variable amount of white on secondaries

Black-necked Grebe

- White panel on secondaries and inner half of primaries, slightly more extensive than in Slavonian Grebe
- No white shoulder-patch (c.f. Slavonian, Red-necked, Great Crested)
- Slightly thinner wings than Slavonian's

Slavonian Grebe

- Neat white secondary panel
- White shoulder-patch (can be missing)

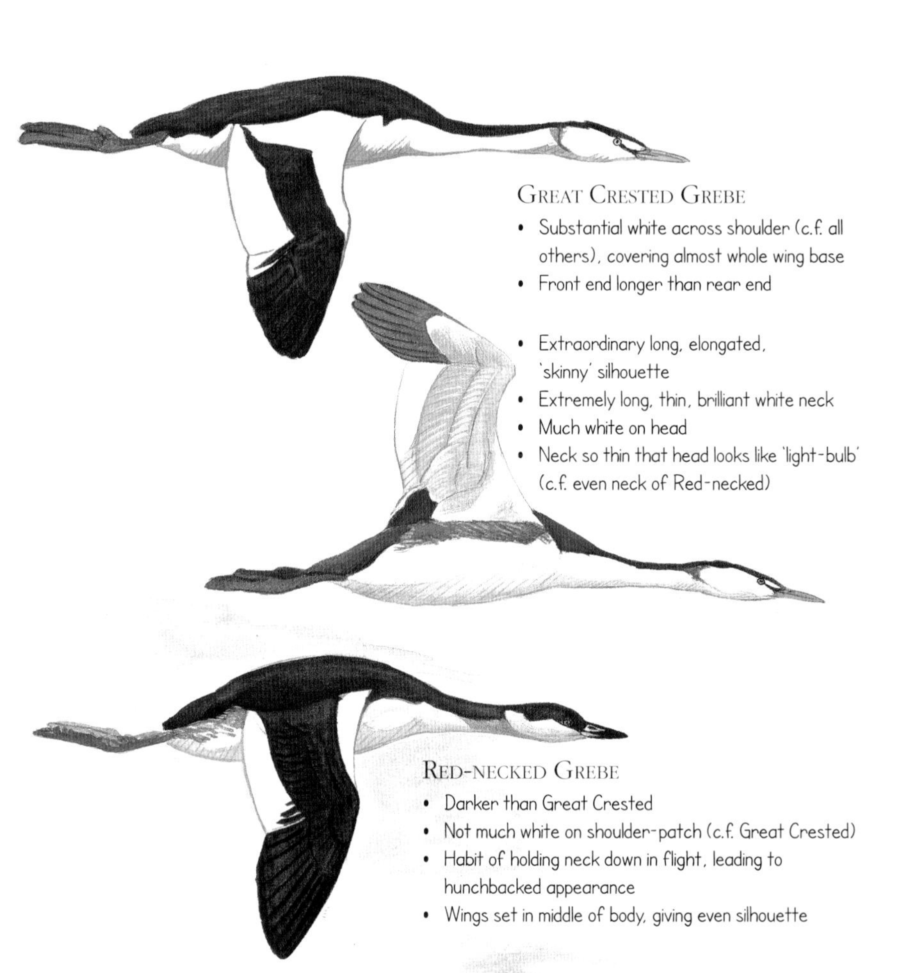

Great Crested Grebe

- Substantial white across shoulder (c.f. all others), covering almost whole wing base
- Front end longer than rear end

- Extraordinary long, elongated, 'skinny' silhouette
- Extremely long, thin, brilliant white neck
- Much white on head
- Neck so thin that head looks like 'light-bulb' (c.f. even neck of Red-necked)

Red-necked Grebe

- Darker than Great Crested
- Not much white on shoulder-patch (c.f. Great Crested)
- Habit of holding neck down in flight, leading to hunchbacked appearance
- Wings set in middle of body, giving even silhouette

Slavonian Grebe

- Neatly defined white wing-panel
- Looks compact
- Short rear end - wings towards back

Larger grebes

Great Crested Grebe (breeding)

- Bill pink, as in winter

Great Crested Grebe (juvenile)

- Black stripes on head (c.f. adult); can be seen from summer right into mid-winter
- Looks as though it is 'wearing striped pyjamas'

Red-necked Grebe

- Larger and sturdier than Slavonian and Black-necked - size closer to that of Great Crested, but plumage tends to be closer to that of the smaller grebes
- Dusky cheeks and neck (c.f. Great Crested)
- Thick-based, dagger-like bill with variable amount of yellow at base
- Dark eye (red in Great Crested, Slavonian and Black-necked)

Great Crested Grebe (neck hunched)

- Spear-shaped pink bill; very long and sharp looking
- Often seen like this, and also resting head and neck on back

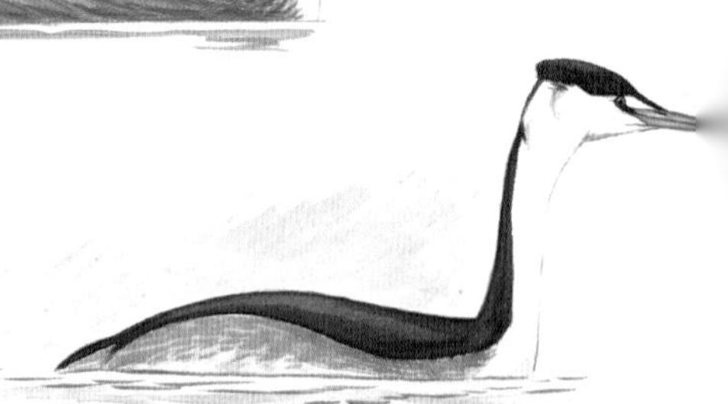

Great Crested Grebe (neck extended)

- Lanky shape, with long, thin neck and long body
- Strikingly white-looking on head, neck and breast
- Black on crown reduced to narrow cap
- Only grebe with white between cap and eye

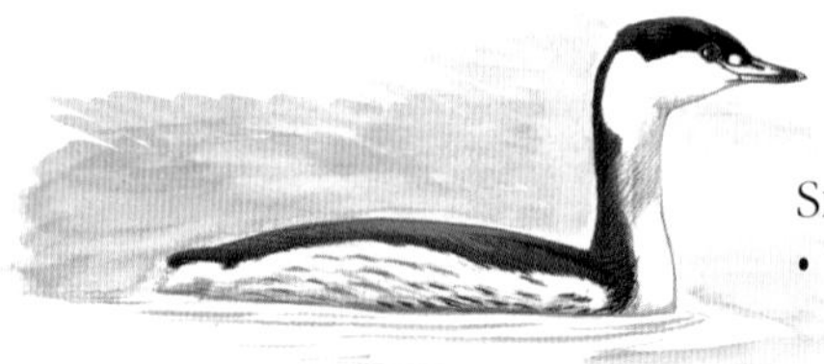

Slavonian Grebe (for comparison)

- In alert posture looks like mini Great Crested in shape - note long neck

GENERAL NOTES ON BIRDS OF PREY

INTRODUCTION The purpose of this section is to compare the shapes and patterns of different birds of prey, related or not. More detailed comparisons of similar species follow in the various sections.

Raptor silhouettes from below

Common Buzzard

- Small head
- Broad wings
- Short tail (about as long as wings are broad front to back)

Northern Goshawk (female)

- As big as Common Buzzard
- Longer tail than Common Buzzard's
- More protruding head than that of Common Buzzard or Eurasian Sparrowhawk
- Bulging secondaries

Northern Goshawk (male)

- Much smaller than female, but still larger than female Eurasian Sparrowhawk
- Longer tail than Common Buzzard's
- More pointed wings than those of Common Buzzard or Eurasian Sparrowhawk

Eurasian Sparrowhawk (female)

- T-shaped outline, with blunt wing-tips but long tail
- Tail has narrow base and square end
- Much blunter wing-tips than any falcon (e.g. Common Kestrel, Peregrine Falcon)

Eurasian Sparrowhawk (male)

- No larger than Common Kestrel
- Sharply pointed corners to tail

Raptor silhouettes from the side

Eurasian Sparrowhawk

- Small raptor
- Blunt wing-tips

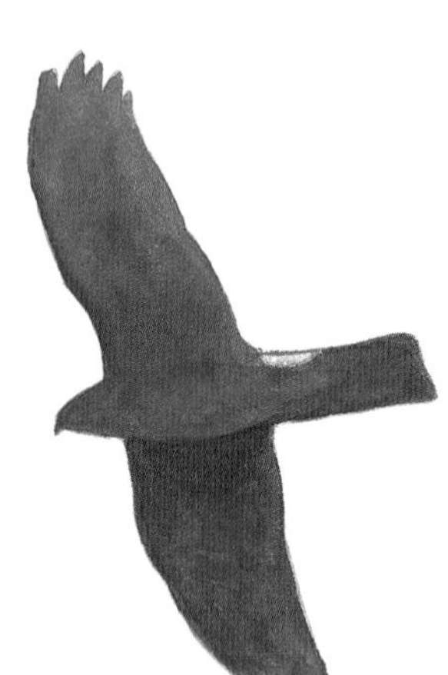

Northern Goshawk

- Medium-sized raptor
- Head projects more than Common Buzzard or Eurasian Sparrowhawk's

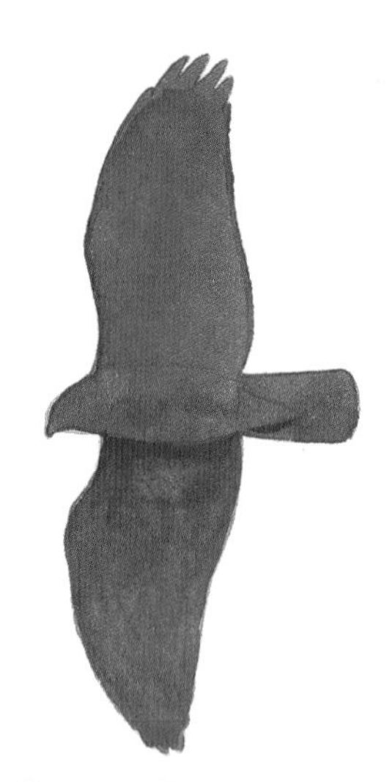

Common Buzzard

- Medium-sized raptor
- Broad wings
- Short tail

Common Kestrel

- Longish tail
- Relatively long arm and hand

Hobby

- Sharp, pointed wings and tail
- Narrow bases to wings
- Short arm, very long hand

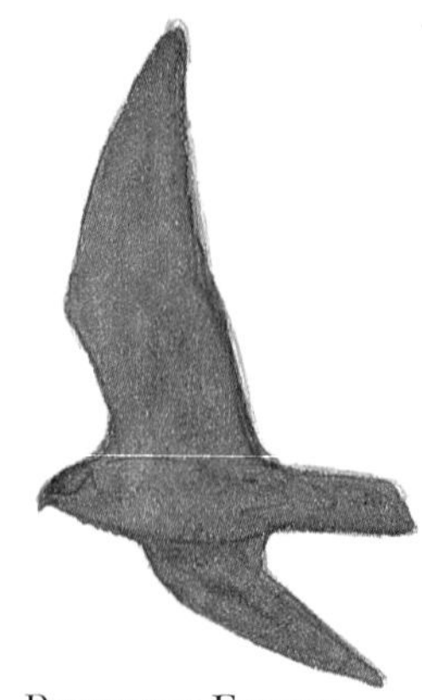

Peregrine Falcon

- Plump and heavily built
- Long, pointed hand with sharp wing-tips
- Broad bases to wings

General raptor ID tips

Eurasian Sparrowhawk (male, soaring)

- Long tail
- Blunt wing-tips
- Tail with a few bars

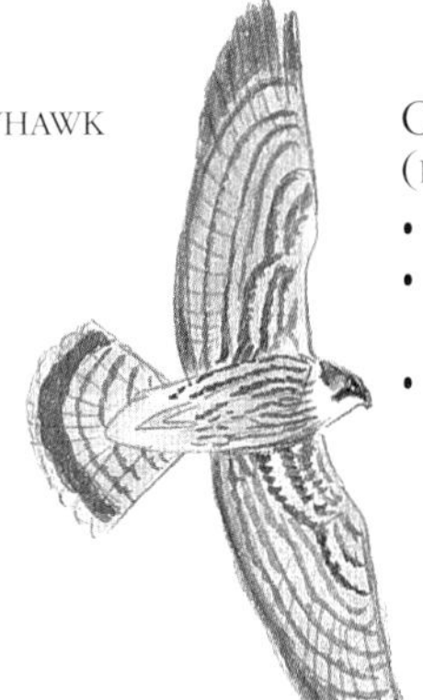

Common Kestrel (female, soaring)

- Small and lightweight raptor
- More pointed wings than Eurasian Sparrowhawk's
- Broad band on tail

Hen Harrier (female)

- Long tail
- Long wings, usually held up in a 'V'
- Slim body compared with other raptors
- Relatively pointed wings

Common Kestrel (male, perched)

- Grey head
- Moustachial stripe
- Rich brick-red upperparts with copious black spots and speckles
- Tail grey with single broad black tip (c.f. female Common Kestrel)

Marsh Harrier (female)

- When soaring, holds wings in shallow but very clear 'V'
- More rounded tail than in kites
- Less obvious 'fingers' than in kites

Black Kite

- Medium-sized raptor, slightly larger than Common Buzzard
- Long tail with sharp-cornered fork (more pronounced in Red Kite)
- Holds wings pressed forwards at carpals, and wings arched down (never up as harriers or Common Buzzard)

Honey Buzzard (dark juvenile)

- Medium-sized raptor similar in profile to Common Buzzard
- Well-protruding head (pigeon-like)
- Inner wing is broadest part; wing pinches in at body

Common Buzzard (dark adult)

- Common medium-sized raptor
- Short tail (no longer than width of wings)
- Rather small head
- Distinctive quick, stiff wingbeats - when soaring holds wings up in a 'V' (c.f. Honey Buzzard, Booted Eagle, Black Kite)

Booted Eagle (dark morph)

- Rather square wings, with fuller hand than a buzzard's
- Translucent inner primaries
- Square tip to tail
- Pale uppertail-coverts

Rough-legged Buzzard (juvenile)

- White base to tail (as immature Golden Eagle)
- This and Common Buzzard have much shorter wings and (especially) tail than Golden Eagle

White-tailed Eagle (juvenile)

- Unmistakable profile, with massive bill, strongly protruding head, broad wings and short tail
- Tail wedge shaped

Golden Eagle (1st winter)

- White patches on underwing (variable) are unmistakable
- White tail with black tip
- Strongly 'fingered' wing-tips

Golden Eagle (adult)

- Much larger than a buzzard
- Very long wings and tail, much longer than buzzard's
- Head protrudes further than buzzard's
- Imperious flight, with slow, deep, regal wingbeats - soars with wings held up in shallow 'V', like Common and Rough-legged Buzzards

HARRIERS

INTRODUCTION Western Marsh Harrier (*Circus aeruginosus*) [L 48cm, WS 126cm] widespread and locally common in larger reedbeds and farmland with wet ditches; in Scandinavia only in extreme south. March–September, but some winter. Hen Harrier (*Circus cyaneus*) [L 48cm, WS 114cm] local on moorland and bog, spreading out in winter to estuaries, farmland, heaths, moors and other open habitats. Montagu's Harrier (*Circus pygargus*) [L 41cm, WS 109cm] summer visitor late April–September, grasslands and larger cultivated fields. Rare in Britain and south Scandinavia. Pallid Harrier (*Circus macrourus*) [L 44cm, WS 110cm] rare visitor from further east.

Harriers are large, quite distinctive birds of prey that specialize in 'quartering', flying low over the ground at slow pace, alternating flaps with glides. When quartering and gliding harriers hold their wings up in a characteristic shallow 'V' shape.

Adult male harriers

Hen Harrier

- Broad, solid black wing-tips
- Head and neck pale smoky-grey
- Belly white, clean and smart
- Underside of wing with distinct dark trailing edge, but otherwise entirely grey

- Some, but rather little contrast between hues of grey on coverts and secondaries respectively
- Trailing edge of upperwing not as distinct as on underwing
- Broader white rump than in male Montagu's

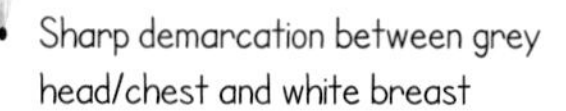

Montagu's Harrier

- Darker grey on upperside than Hen
- Thin chestnut streaks on belly
- Chestnut and black bars on underwing-coverts - underwing looks a bit messy
- Slightly longer and thinner looking tail than Hen's, with bars at edge

- Two shades of grey on upperwing
- Wing-tip much more pointed than relatively blunt tip of Hen
- Black secondary bar diagnostic of Montagu's (both sexes)
- White rump on Montagu's slightly smaller than that of Hen and less noticeable

---

Marsh Harrier

- Tricoloured upperwing (black, brown, grey)
- Leading edge of wing often creamy white (as female Marsh)
- Dark chestnut-tinted breast unique among male harriers
- Clean, smart underwing reminiscent of Hen's, but wing broader, coverts brown tinged and complete lack of dark trailing edge

- Unmarked grey tail
- Pale, lightly streaked head
- Distinctive rust-brown underparts with thickish, dark-brown streaks

- Smaller than female
- Black wing-tips (as all male harriers)
- Pale grey underwing (in younger males, browner)
- Paler breast than female's

Female and immature Marsh Harrier

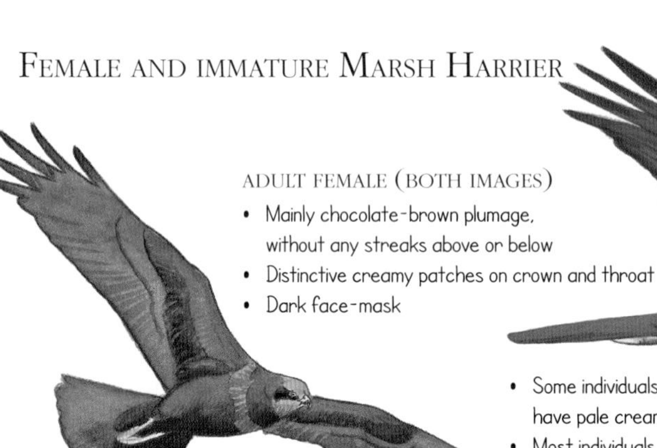

ADULT FEMALE (BOTH IMAGES)

- Mainly chocolate-brown plumage, without any streaks above or below
- Distinctive creamy patches on crown and throat
- Dark face-mask
- Some individuals (not all) have pale creamy breast-band
- Most individuals have creamy front edge to wing
- Tail paler brown than rest of bird (see immature)

IMMATURE (BOTH IMAGES)

- Generally darker than adult female (and some simply dark brown all over)
- Brighter yellow markings on head than adult female's
- Always lacks pale breast-band
- Narrow pale bar between coverts and main flight feathers indicative of juvenile
- Usually lacks any cream colour on leading edge

3RD-YEAR MALE

- Still shows creamy crown and dark face-mask
- Tail plain (not streaked), but has dark subterminal band
- Greater contrast between coverts and flight feathers than in adult male
- Dark trailing edge to wing

HEN HARRIER (JUVENILE FEMALE)

- Slimmer than Marsh, with slightly longer tai
- Narrower wing-tips than Marsh's
- Wings heavily barred underneath

Perched harriers

Montagu's Harrier (female)

- Smaller head than Hen's, with more obvious facial pattern
- Broad white supercilium
- Clear white half eye-ring below eye (c.f. female Hen)
- Cheeks very dark brown and contrasting

Hen Harrier (female)

- Strong, bold stripes down pale breast
- Narrow supercilium and small, indistinct patches around eye (c.f. Montagu's and juvenile Hen)
- Obvious pale neck ring (c.f. Montagu's)

Marsh Harrier (male)

- Pale head contrasts with rest of body
- Mainly brown body (other male harriers grey)
- Creamy-buff patch on shoulders

Marsh Harrier (female)

- Distinctive creamy patches on head
- Brown tail without bars
- Larger and heavier bodied than male

Adult female 'ringtail' harriers

Montagu's Harrier (adult female)

- Narrow wings with sharper tips than Hen's
- Dark band on upperwing (c.f. Hen)
- Underwing-coverts ginger coloured
- More obvious barring on underwing-coverts than Hen's

Hen Harrier (adult female)

- Broader, blunter wings than Montagu's, with 'bulging arm'
- Large white rump-patch (c.f. Montagu's)
- Black 'rings' on tail (c.f. female Marsh)
- Underparts dark streaked on pale (c.f. Marsh) background

JUVENILE 'RINGTAIL' HARRIERS

MONTAGU'S HARRIER

- Rich reddish-brown tone to underparts easy distinction from juvenile Hen (and all other harriers except juvenile Pallid - see below)
- Underwing-coverts equally richly coloured, and lack barring (c.f. adult Montagu's; all Hen Harriers)
- No pale collar around neck

HEN HARRIER

- Pale tips to greater wing-coverts make narrow bar on upperwing (c.f. adult female harriers)
- Dark wing-tips and bars not as boldly defined as those in female Montagu's
- Breast streaks obvious (c.f. juvenile Montagu's)
- Whitish neck-ring

MONTAGU'S HARRIER

- Broad supercilium
- White almost encircles eye
- Dark spot on ear-coverts

MONTAGU'S HARRIER

- Ear-covert 'spot'
- No collar
- More obvious black tip to outer primaries

PALLID HARRIER

- Complete pale collar
- Brown on ear-coverts reaches gape to complete mask around eye
- Dark side of neck
- Slightly less dark on wing-tips than in Montagu's

BUZZARDS

INTRODUCTION Common Buzzard (*Buteo buteo*) [L 53cm, WS 120cm] very common and widespread (summer visitor to Scandinavia) in habitats mixing woodland with open country, including farmland, foothills, moorland and wetlands. Rough-legged Buzzard (*Buteo lagopus*) [L 55cm, WS 137cm] occurs right up to the Arctic as a summer visitor to tundra and mountains. In winter retreats to the Low Countries eastwards to Central Europe; rare Britain. European Honey Buzzard (*Pernis apivorus*) [L 55cm, WS 120cm] more of a forest bird than Common Buzzard, requiring woodland glades rather than open country. Widespread summer visitor May–September, wintering in Africa. Rare in Britain, the Low Countries and Norway.

When flying straight, Common Buzzard has noticeably stiff wingbeats alternating with short glides; thus the flight lacks the more imperious, more controlled flying of Honey and Rough-legged Buzzards. The others have a much more fluid flight style, with slower wingbeats; the emphasis is on the upstroke. When soaring upwards, Common and Rough-legged Buzzards typically hold their wings up at a slight angle. Honey Buzzards tend to hold their wings level, or arched very slightly down. Both Common and Rough-legged hover, but Honey does not.

Buzzard silhouettes from below

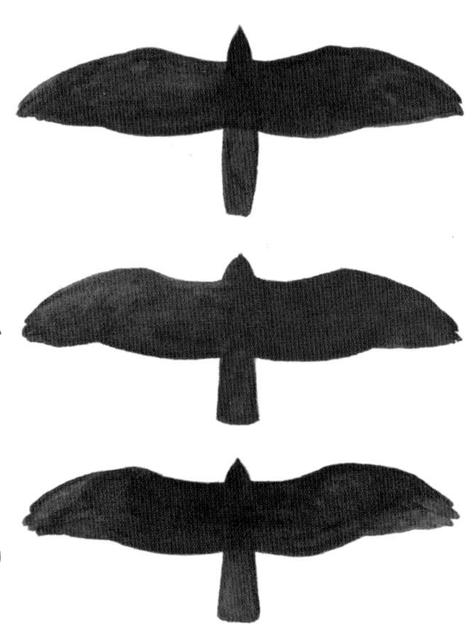

Honey Buzzard

- Longer tail than in Common; longer than width of wing from front to back
- Slender, protruding head (c.f. Common Buzzard), a bit pigeon- or Cuckoo-like
- Wings broadest in middle - relatively pointed at tips and visibly narrowing towards body (quite distinctive at long range)
- Tail sides slightly convex

Common Buzzard

- Tail as long as width of wing (front to back) or shorter
- Thick neck and short, rounded head
- Compact

Rough-legged Buzzard

- Longer wings than Common's, leading to quite distinctive shape
- More obvious forward kink at carpal joints when gliding
- Secondaries bulge more
- Longer tail than Common's

Buzzard silhouettes

Rough-legged Buzzard

- When gliding, raises arm, but hand is fairly flat, making obvious kink halfway along wing - Common (and Honey) never shows this

Common Buzzard

- In contrast to many similarly sized species (e.g. Honey Buzzard, kites) holds wing up at slight angle when soaring (i.e. gaining height while circling)
- When gliding (i.e. moving forwards without flapping wings and not gaining height) tends to hold wings flat

Common Buzzard (3 in air, typical profile)

- Wings held pressed slightly forwards
- Not much of kink at carpals (c.f. Rough-legged)
- Rather small head
- Compact appearance

Common Buzzard: adult from juvenile

Adult

- Obvious dark tail-band
- Well-defined dark trailing edge to wings

Juvenile

- Outer tail-band as narrow and poorly defined as other bands on tail
- Often more obvious streaks down breast than in adult

Common Buzzard variation: colour morphs and ages

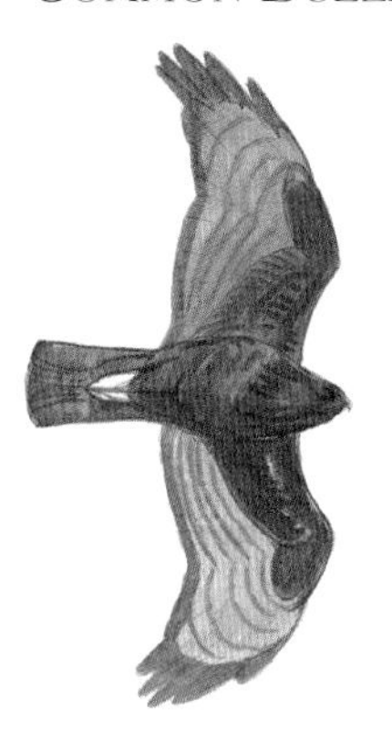

DARK JUVENILE

- Tail with many fine bars
- Heavily barred on paler flight feathers

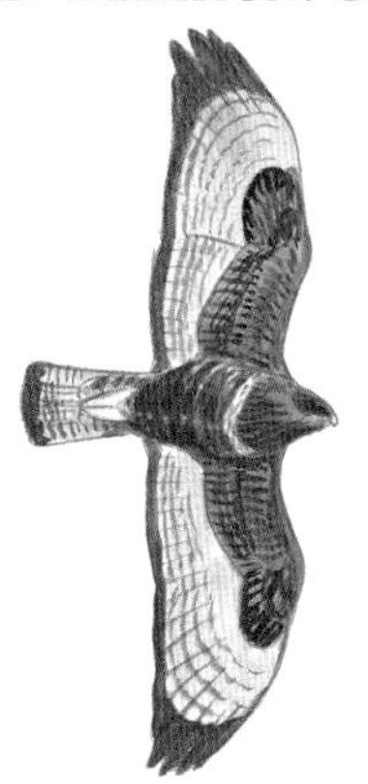

TYPICAL DARK ADULT

- Carpal patches well marked
- Well-marked, broad tail-band

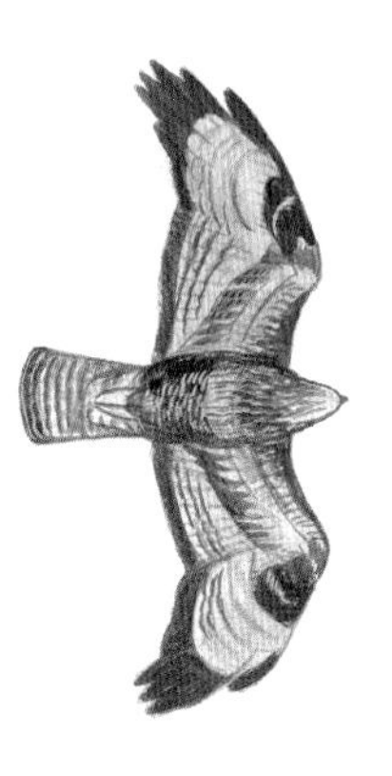

MOTTLED JUVENILE

- Carpals on this bird conspicuously dark, making it look similar to Rough-legged
- Base of tail barred and pale brown

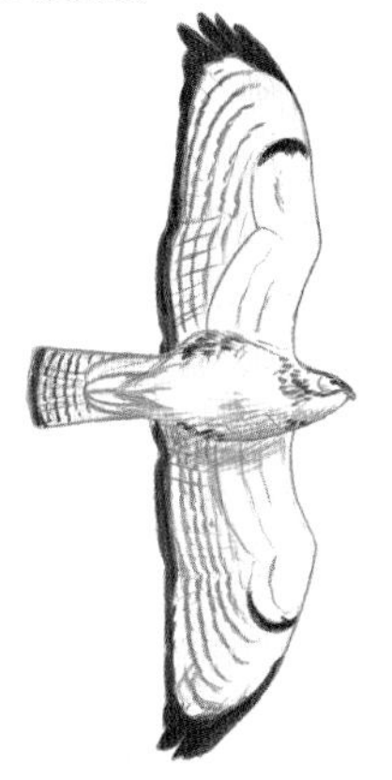

PALE JUVENILE

- Most confusing of patterns, but quite common - easily confused with various other raptor species

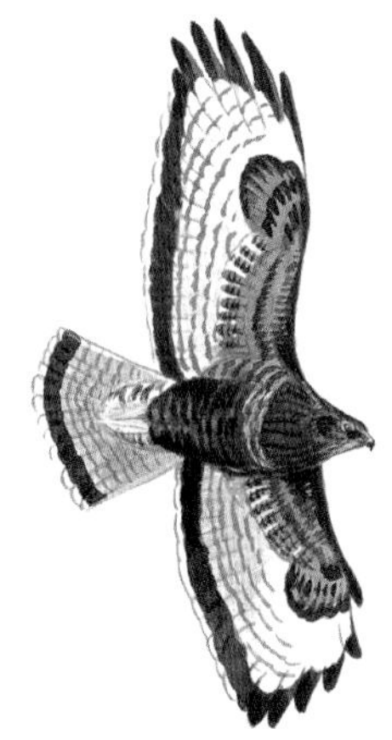

TYPICAL DARK ADULT

- Typically pale band ('medallion') across chest
- Densely barred grey tail

MOTTLED ADULT

- Barred underwings between coverts and trailing edge (5-7 bars, making underwing appear greyish)
- Coverts typically mainly dark, contrasting with bases of primaries

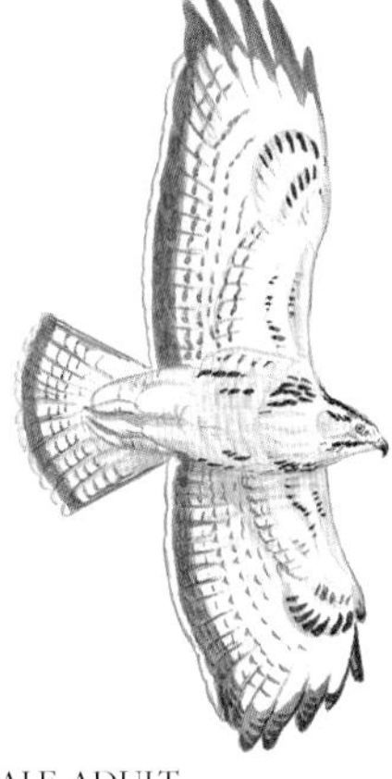

PALE ADULT

- Sparsely streaked below
- Dark trailing edge to wings
- Dark tail-band

Ageing Rough-legged Buzzard

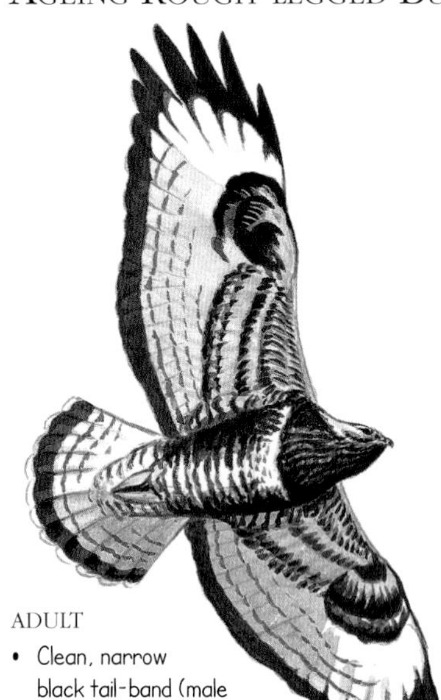

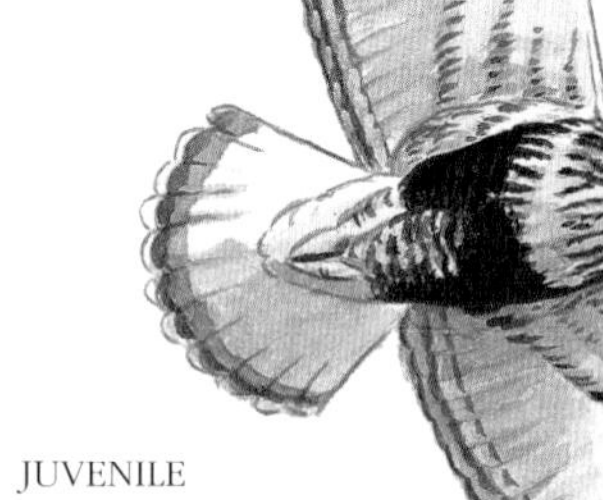

ADULT

- Clean, narrow black tail-band (male may have some narrow bands inside main band)
- Black carpal patches usually much more pronounced than in Common
- Noticeably pale underwing
- Well-defined dark trailing edge to wing

JUVENILE

- Weak tail-band, although broader than adult's
- Often very dark belly, which is darker than throat
- Clean pale uppertail-coverts
- Carpal patches often more obvious than in adult

ADULT

- Black tail-band
- A few extra bands - this bird is probably male
- Usually darker brown above than Common
- Pale base to primaries - good distinction from Common

JUVENILE

- Broad dark tail-band (without extra bands) best distinction from adult - takes up nearly half of tail

Rough-legged Buzzard vs Common Buzzard

Rough-legged Buzzard

Juvenile

- More boldly patterned and contrasting than Common
- Bold black carpal patches larger than in Common, and usually more conspicuous
- Streaks on belly (usually absent in Common)
- Juveniles often exhibit warm buff colour on belly and coverts (as here)

Immature

- Huge carpal patches
- Whitish, lightly patterned coverts
- Whitish leading edge to wings (c.f. Common)

Adult

- Dark belly patch lacking in Common (most obvious in female Rough-legged, and can be almost missing in male)
- Sharply defined tail-band indicates adult; several smaller bands suggest a male (in female, just 1-2 extra bands, which tend to merge into main tail-band)
- Tail has white base (diagnostic - but note possibility of confusion with very pale Common Buzzard)

Common Buzzard

Pale morph

- Carpal patch no more than 'comma' - always much more extensive in Rough-legged
- Whitish base to tail does not contrast with darkish belly, as it does in Rough-legged

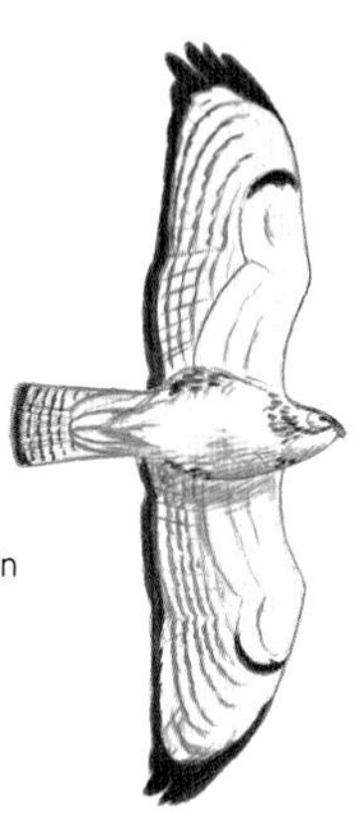

Dark morph

- Carpal patches well marked, but do not stand out as clearly as Rough-legged's
- Base of tail clearly marked with bars (absent in Rough-legged)
- Lots of barring on inner half of flight feathers

Typical morph

- Pale band on belly - present in many individuals, and never as obvious in Rough-legged
- Underparts dark, without obvious patch on belly

Ageing and sexing Honey Buzzard

TYPICAL ADULT MALE

- Distinctly grey head
- Whiter-looking outer wings than female
- Long 'step' from tip to next tail-bar

PALER ADULT MALE

- Pale eyes (dark in adult Common)
- Pale grey head
- Three tail-bands: two at base and one at tip (a little like a DNA printout)
- Dark trailing edge to wing (c.f. juvenile Honey Buzzard, Common Buzzard)

ADULT FEMALE

- Closer barring on breast than in male
- Slightly more black on wing-tips than in male

PALE JUVENILE

- More easily confused with Common Buzzards than Honey Buzzard adults; darker on underside and tends to have shorter tail and wings than adults, all characteristics that are closer to Common
- Only a few bars on tail
- Always lacks pale band across chest
- A few quite clearly defined bands across primaries and secondaries (standard distinction from adult Common, but note possible confusion with juvenile Common)

DARK JUVENILE

- Duskier below than adult
- More bars on wings than in adult
- Lacks black trailing edges to wings
- Extensive black on wing-tips (similar in Common)

Common Buzzard vs Honey Buzzard

Common Buzzard

PALE MORPH

- Many bars on tail
- Breast streaking can be very similar to that in pale juvenile Honey

TYPICAL MORPH

- Not such clear bands across secondaries and primaries as in Honey
- Plumper body than that of slimline Honey

DARK MORPH

- Short, rounded head

Honey Buzzard

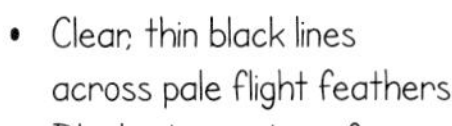

ADULT MALE

- Clear, thin black lines across pale flight feathers
- Black at very tips of wings
- Grey head with yellow eye
- Oval-shaped black carpal patch

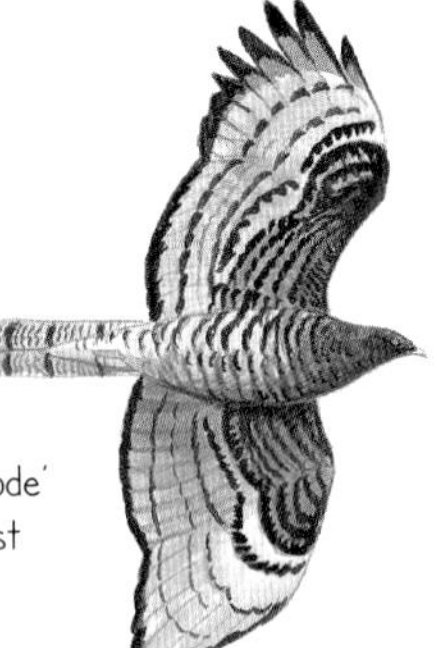

ADULT FEMALE

- Long tail with 'barcode'
- Barred across chest

DARK JUVENILE

- Dark eyes like Common's, but not adult Honey's
- Slit at tip of tail
- Clearer lines across flight feathers than in any Common

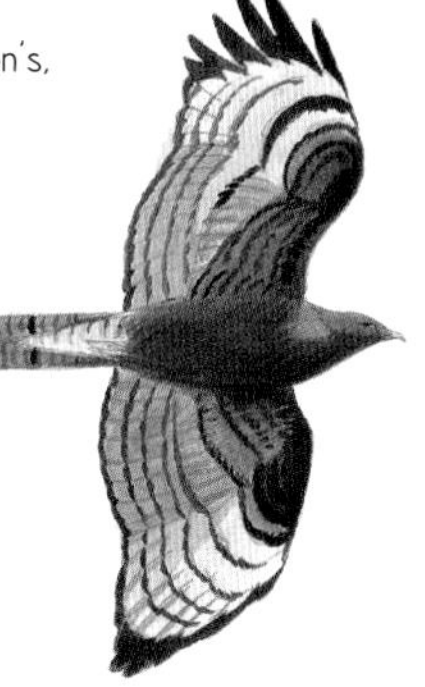

Perched buzzards

Common Buzzard

ADULT

- Dark, robust raptor
- Rounded head, short neck, hunched shoulders
- Distinctive pale band across chest of most individuals
- Yellow bill with black tip
- Tail projects beyond wing when perched

PALE JUVENILE

- Often has pale head - possible confusion with Rough-legged, Marsh Harrier and Osprey
- Any oddly coloured, creamy bird of this size likely to be Common Buzzard
- Usually whitish all down front
- Bars on tail; inner half of tail sometimes white, but never as cleanly defined as Rough-legged's

TYPICAL JUVENILE

- Paler brown than adult
- Distinctly paler below than adult
- Coarsely streaked underparts

Rough-legged Buzzard

JUVENILE

- Much paler head and neck than in most Commons, but beware pale juvenile of latter
- Blackish belly (c.f. adult Rough-legged)
- Narrow dark eye-stripe more marked than in pale Common

ADULT

- Heavy barring on breast on pale background
- White upper half of tail usually clearly visible

EAGLES

INTRODUCTION Golden Eagle (*Aquila chrysaetos*) [L 86cm, WS 202cm] localized resident in Scotland (a few in northern England) and Scandinavia in remote mountainous regions, wandering to lowlands. White-tailed Eagle (*Haliaeetus albicilla*) [L 87cm, WS 220cm] rare resident on coasts of Scotland, Scandinavia and the Baltic region, occurring in a variety of habitats including cliffs and lakes; rare winter visitor elsewhere. Booted Eagle (*Aquila pennata*) [L 46cm, WS 125cm] scarce summer visitor March–September, to central and southern France, occurring in scrubby countryside with open woods, usually in the hills. Short-toed Eagle (*Circaetus gallicus*) [L 61cm, WS 185cm] rare summer visitor March–September, to France, in mountain ranges with sunny, scrubby, rocky slopes.

Eagle characteristics include large size (although Booted is not actually much bigger than a buzzard); protruding head; broad wings with 'fingered' tips; powerful, 'regal' flight characterized by very slow, steady wingbeats with long glides (shorter in White-tailed); and ability to soar.

White-tailed Eagle

Golden Eagle

SILHOUETTES

- Only eagle to soar with wings upswept in shallow 'V'
- Relatively long tail

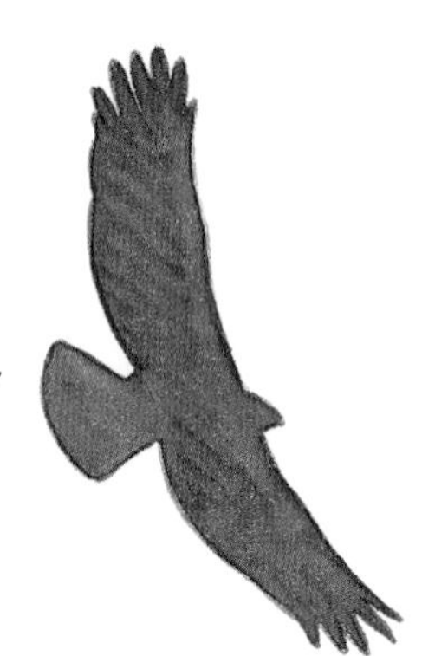

- Round-tipped tail
- Wings 'pinched in' at base, giving distinctive 'S'-shaped trailing edge
- Quite rounded wing-tips

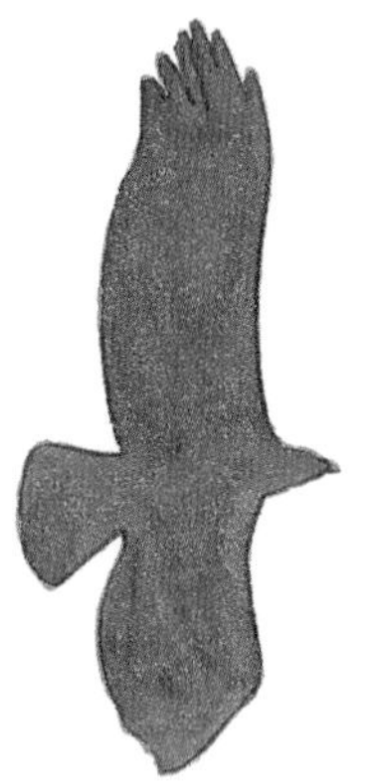

SUBADULT

- Diffuse tail-band suggests this bird is more than three years old
- Smaller white patches on wing than in young birds
- Long, rectangular wings with clear 'fingers'
- White base to tail

- Takes 5-7 years to acquire adult plumage, getting progressively darker and losing all white markings
- Very long wings (c.f. buzzards), distinctively rectangular in shape
- White tail-base and wing-patches diagnostic of Golden immature (first three years)

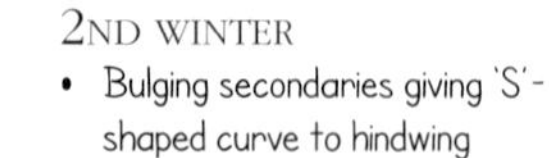

2ND WINTER

- Bulging secondaries giving 'S'-shaped curve to hindwing
- Bright white base to tail (above and below) with dark tip (no other eagle shows this)
- White square in middle of wing (unique)

ADULT

- Golden 'shawl' on back of head
- Feathered legs (look like trousers) - c.f. White-tailed

Golden Eagle (adult)

- Looks dark at distance, with few markings - can be helpful in identification
- Golden band across upper wing (varies)
- Greyish tail barred with black tip
- Wing-coverts greyish with black bands (c.f. immature)

Short-toed Eagle

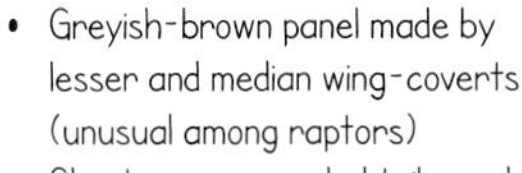

- Greyish-brown panel made by lesser and median wing-coverts (unusual among raptors)
- Short, square-ended tail, much shorter than Golden's
- Whitish rump (adult, not juvenile)
- Dark bars on tail (c.f. Booted)

- Very pale underneath, with variable dark lines of spots, sometimes quite heavily marked); from distance just looks whitish (unique)
- Dark hood, usually sharply demarcated from pale lower breast and belly
- Square-ended tail with sharp corners
- Very long and broad wings (distinctive)

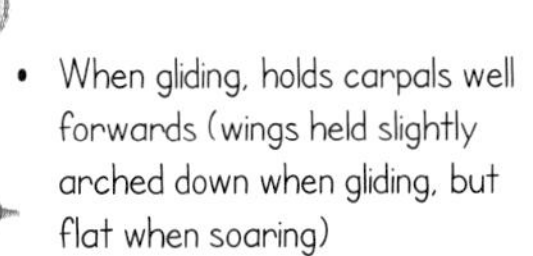

- When gliding, holds carpals well forwards (wings held slightly arched down when gliding, but flat when soaring)
- No sign of dark carpal patches (rules out Osprey and pale Common and Honey Buzzards)
- A few (3-4) evenly spaced dark bars on tail

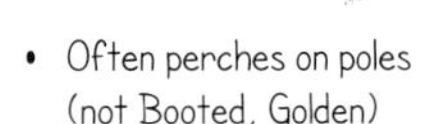

- Often perches on poles (not Booted, Golden)
- Disproportionately large, broad head

Booted Eagle

- Appears in two main colour morphs. Most birds are pale and easy to identify, but others are darker and can resemble buzzards.
- Small for an eagle; size as Common Buzzard
- Conspicuous (and curious) white 'landing lights' at front of wing where it meets body (not on all birds)
- Distinctive pale band across upper wing (as Black Kite, but different from Common Buzzard)
- Pale uppertail-coverts
- Pale crown and dark face distinctive

PALE MORPH

- Dark flight feathers, contrasting whitish coverts

INTERMEDIATE MORPH

- Square tip to tail
- Dark central wing band
- Pale uppertail-coverts (not visible here)

DARK MORPH

- Rather square wings, with fuller hand than buzzard's
- Translucent inner primaries - all morphs

GOSHAWK & SPARROWHAWK

INTRODUCTION Both species widespread residents. Northern Goshawk (*Accipiter gentilis*) [L 61cm, WS 109cm] resident in mature forests of all kinds (including boreal), sometimes breeding in small tracts of woodland and regularly hunting over open country; scarce and localized in Britain. Eurasian Sparrowhawk (*Accipiter nisus*) [L 33cm, WS 68cm] breeds in woods but occurs in all kinds of habitats, even gardens; common in Britain.

The difference in size between the sexes of both species is substantial, with females being bigger and much heavier than males. Male Sparrowhawk is as small as a Common Kestrel, while female Goshawk is as big as a Common Buzzard; more awkwardly, female Sparrowhawk is similar in size to male Goshawk, and this is where identification is most problematic.

SPARROWHAWK (FEMALE)

- Shorter inner wing (arm) than Goshawk's, with less pointed outer wing (hand)
- Face-mask not as bold as Goshawk's
- Thin legs like 'knitting needles'; Goshawk's thick and powerful

SPARROWHAWK (MALE)

- Orange-barred underparts diagnostic

GOSHAWK

- Typical view in hunting mode – in common with Sparrowhawk, tries to ambush prey by hiding out of sight, then accelerating towards it, as this bird is doing

Sparrowhawk (female)

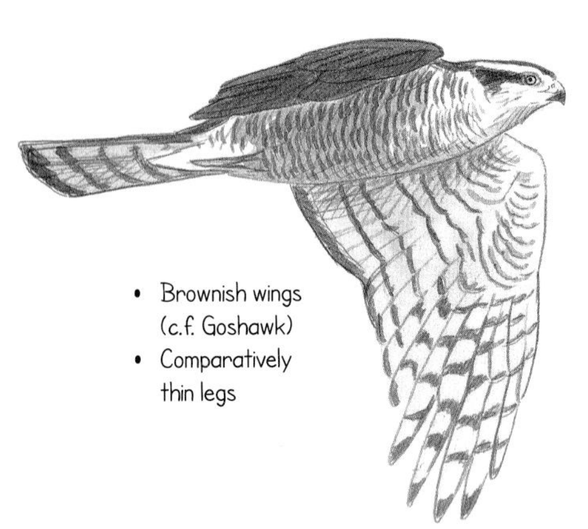

- Brownish wings (c.f. Goshawk)
- Comparatively thin legs

- Often fans tail when soaring, as does Goshawk

Goshawk

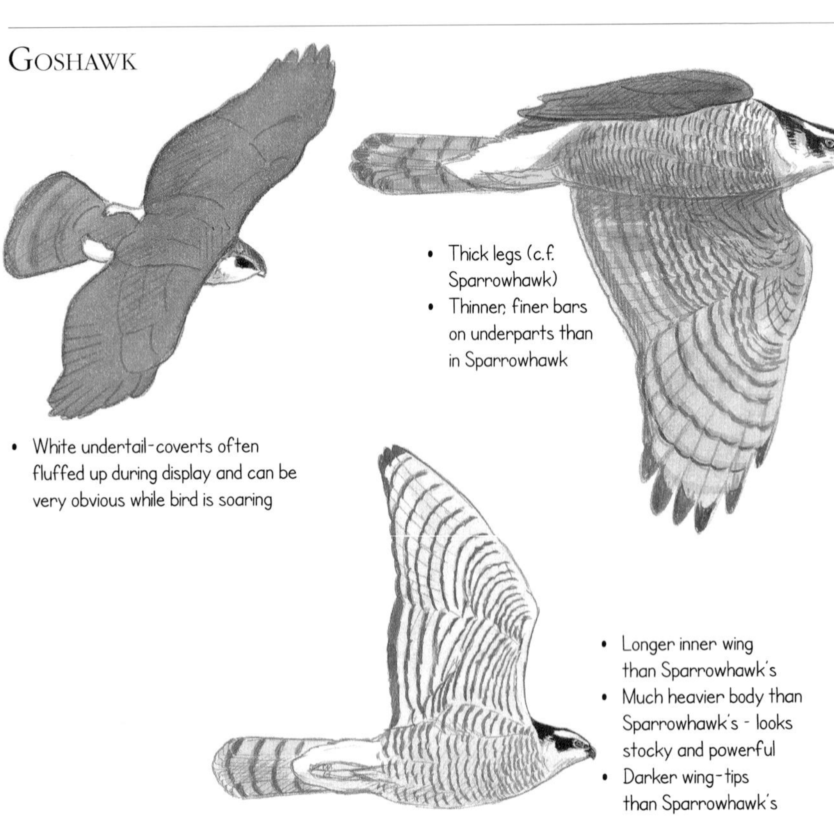

- Thick legs (c.f. Sparrowhawk)
- Thinner, finer bars on underparts than in Sparrowhawk

- White undertail-coverts often fluffed up during display and can be very obvious while bird is soaring

- Longer inner wing than Sparrowhawk's
- Much heavier body than Sparrowhawk's - looks stocky and powerful
- Darker wing-tips than Sparrowhawk's

Sparrowhawk (female)

Goshawk

- Very broad tail-base (c.f. Sparrowhawk)
- Comparatively shorter tail than Sparrowhawk's
- Looks 'hip heavy'
- Longer neck and more protruding head than Sparrowhawk's

SPARROWHAWK & KESTREL

INTRODUCTION These two are often the most common small raptors in an area. Both widespread in region. Common Kestrel is inveterate hoverer; Eurasian Sparrowhawk simply does not hover.

Kestrel (male)

- Grey head
- Tail grey with single broad black tip (c.f. female)
- Spots on buff-coloured breast
- Often perches high up, conspicuously (Sparrowhawk perches hidden)

Kestrel (female)

- Dark primaries contrast with light inner wing (both sexes - good comparison to Sparrowhawk's upperwing, which has no contrast)
- Tail brown with broad dark tip, but also several well-spaced narrow bars (c.f. male)
- Moustache on face (c.f. Sparrowhawk)

Sparrowhawk (male)

- About size of Kestrel, but plumper
- Dark blue-grey on upperparts (c.f. Kestrel, female Sparrowhawk)
- Often shows white 'paint blobs' down back, plus white spot on nape
- Orange-barred breast (unique)

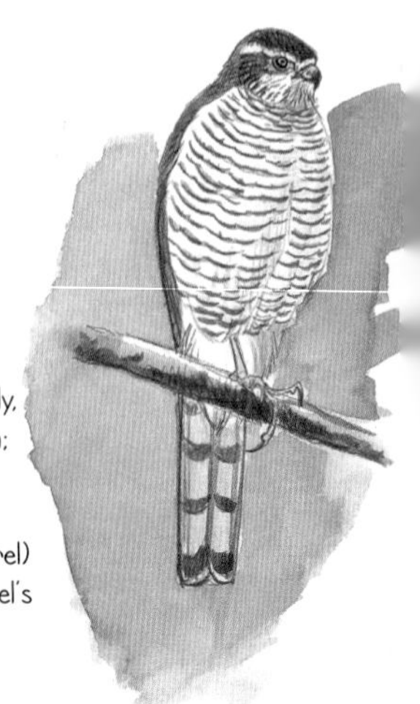

Sparrowhawk (female)

- Almost always perches furtively, amid cover or otherwise hidden; often moves on quickly
- Barred across chest (not streaked down chest, as Kestrel)
- Whiter underparts than Kestrel's (and male Sparrowhawk's)
- White supercilium

SPARROWHAWK (FEMALE)

- Flies with sequences of flaps and glides (Kestrel glides less). Looks powerful

COMMON CUCKOO

- Small head protrudes a long way forwards
- Pointed bill
- Flight characteristic: does not glide, but flaps continuously, and wings barely seem to rise above horizontal. Often 'cuckoos' in flight

KESTREL (FEMALE)

- Flies with shallow flaps
- Rounder tail-tip than Sparrowhawk's
- Wings narrow, long and pointed; held out straighter than Sparrowhawk's

SPARROWHAWK (FEMALE)

- Long, square-ended tail
- Comparatively blunt wing-tips (c.f. Kestrel)
- Broad inner wings

KESTREL (FEMALE)

- One broad and several narrow bands on tail (in male just broad band near tip)
- Looks slim and lightweight

PARROWHAWK (MALE)

Orange underparts

FALCONS

INTRODUCTION Falcons are distinguished from other birds of prey by their relatively narrow, sharply pointed wings. Merlin (*Falco columbarius*) [L 29cm, WS 62cm] breeds on moorland and tundra in northern Britain, Ireland, Iceland and Scandinavia; in winter moves south and ranges widely. Eurasian Hobby (*Falco subbuteo*) [L 32cm, WS 77cm] summer visitor April–October, to the Continent, southern Britain and milder parts of Scandinavia, breeding in various types of open country with scattered trees. Peregrine Falcon (*Falco peregrinus*) [L 45cm, WS 102cm] breeds widely on cliffs and sometimes tall buildings in more rugged country (mountains, sea coasts) throughout the region. Ranges widely in winter. Common Kestrel (*Falco tinnunculus*) [L 34cm, WS 73cm] occurs in a wide variety of open habitats, breeding on ledges and in holes in trees. Widespread.

Perched juveniles

Merlin

- Similar to adult female Merlin, but may be more reddish-brown
- Not as scaly as other juvenile falcons
- Rich reddish-brown barring on breast, or even with spots

Hobby

- Buff breast with bold, well-defined streaks (pattern similar to that in adult Hobby, which has white breast)
- Warm buff undertail-coverts (c.f. adult Hobby)
- Scaly upperparts (c.f. adult Hobby)
- Brownish tinge to upperparts (slate grey in adult)
- Pale patch on nape (not adult)

Peregrine Falcon

- Narrower black moustache than adult's
- Brownish upperparts (grey in adult)
- Bold streaks down breast (in adult, bars go across chest)
- Barred undertail-coverts
- Cere bluish (yellow in adult)

Distant flight silhouettes

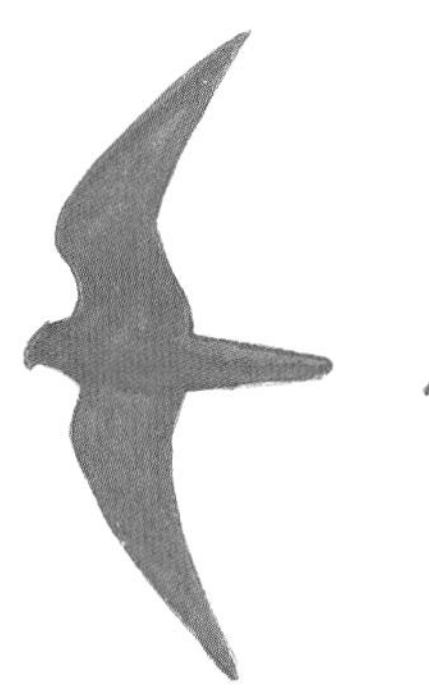

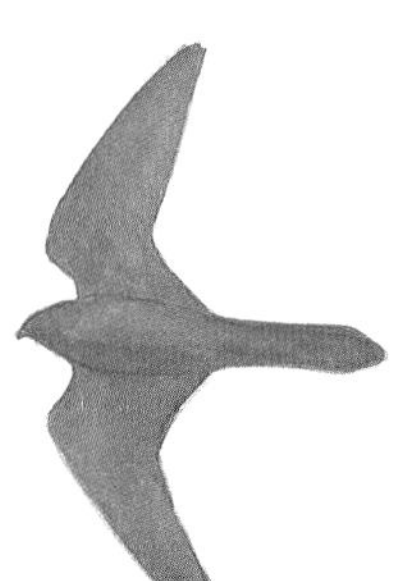

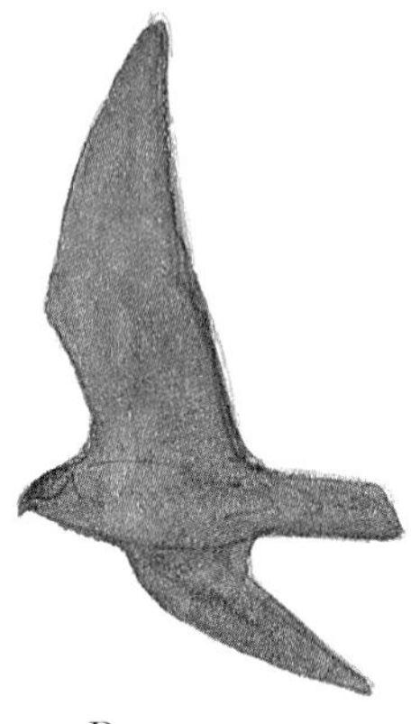

Hobby

- Famous for resembling an oversized swift
- Tail much shorter than Kestrel's
- Very sharply pointed wings

Kestrel

- Blunter wings than Hobby's
- Longer tail than most falcons'

Merlin

- In shape most similar to Peregrine, with broad-based wings with pointed tips.
- Tiny and compact
- Faster wingbeats than those of the other falcons
- Tail square ended

Peregrine

- Powerful, bulky, muscular raptor
- Short arm
- Short tail

Kestrel flight silhouettes

Soaring

When soaring, wing-tips can look blunt rather than pointed

Closed tail

- Much longer tail than in the other falcons

Open tail

- When not hovering, distinctive flight action somewhat lacks power and purposefulness of other falcons; flight loose and rather flappy
- Does not glide much when in normal level flight (c.f. Sparrowhawk)

Peregrine Falcon

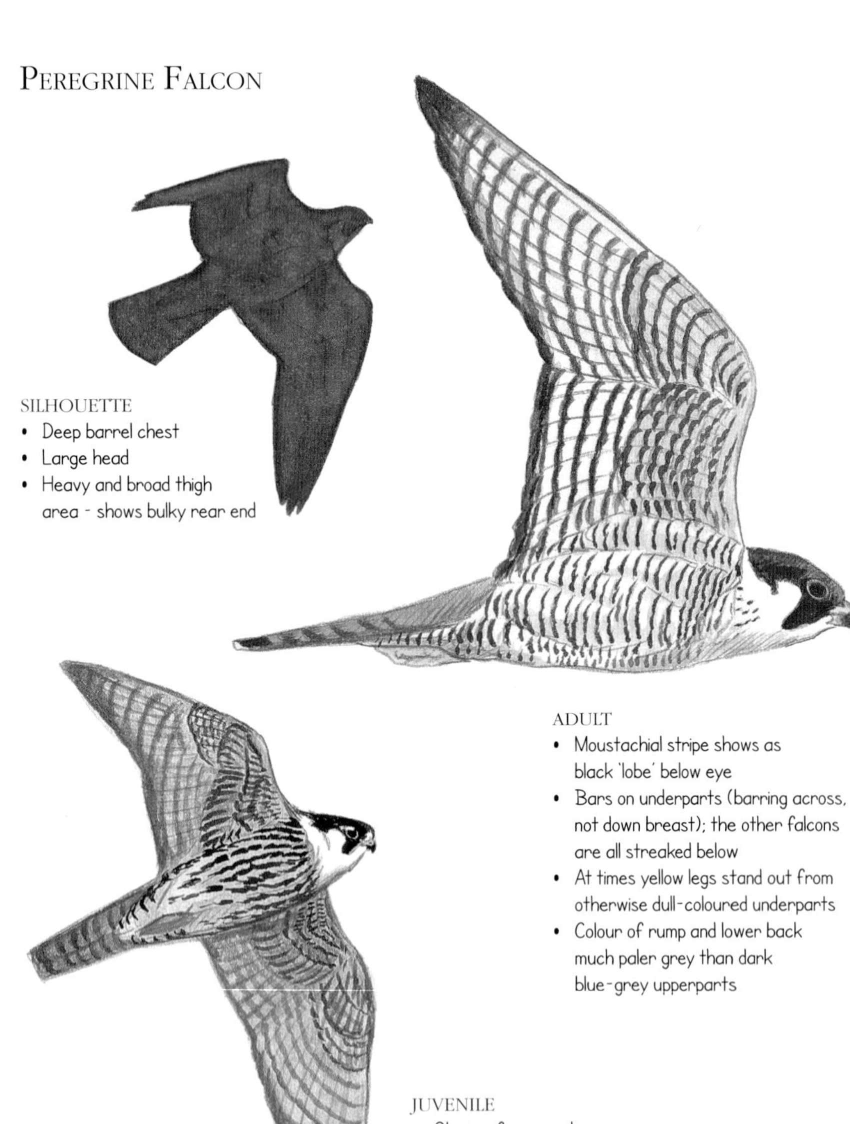

SILHOUETTE
- Deep barrel chest
- Large head
- Heavy and broad thigh area - shows bulky rear end

ADULT
- Moustachial stripe shows as black 'lobe' below eye
- Bars on underparts (barring across, not down breast); the other falcons are all streaked below
- At times yellow legs stand out from otherwise dull-coloured underparts
- Colour of rump and lower back much paler grey than dark blue-grey upperparts

JUVENILE
- Obvious face-mask
- Noticeably 'triangular' wings, very broad based but tapering sharply to pointed tip
- Buff ground colour to underparts
- Streaked rather than barred on breast and belly

• Wings look distinctly triangular
• Square-ended tail
• Note anchor shape
• Usually flies with distinctly stiff and shallow wingbeats, unless engaged in pursuit of prey (then, wingbeats deep and powerful)
• Sometimes fans tail when soaring
• Carpal bends obvious
• Pale below
• Looks very pale grey from below; paler than Hobby
• Broad thighs ('cellulite thighs')
• Broad, short tail
• Broad shouldered and powerfully built
• When flying, sometimes seems to flap only tips of wings, giving impression of effortless power
• Famous for spectacular dives from great height; also chases birds low and close to ground at high speed

Hobby

SILHOUETTE
- Wings long, sharply tapered and pointed (short arm, long hand)
- Wings flexed back
- Narrow at base of tail (c.f. Peregrine)
- Tail only moderately long

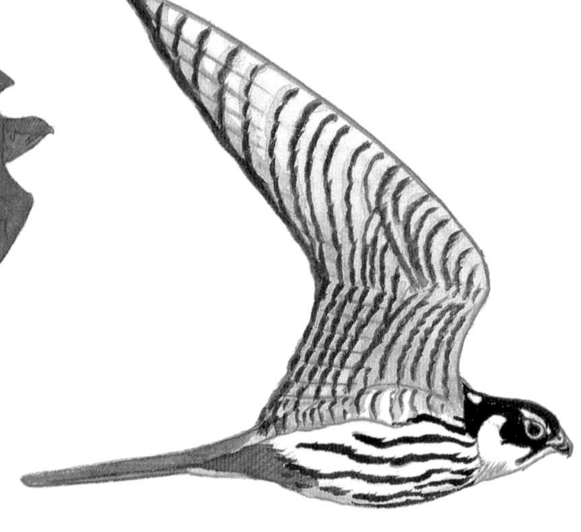

ADULT
- Light buff, rather than white below (exceptional light needed to appreciate this)
- Strongly capped effect, with conspicuous white cheeks
- Strong rusty-red vent and undertail-coverts - in good light can be blazingly obvious

JUVENILE
- Lacks reddish thighs of adult
- Narrow-based wings very sharply pointed

SOARING
- Often glides very slowly high in sky, like a giant, slow-motion swift
- Even when fanned, tail still looks narrow

IN GOOD LIGHT

- Richly striped underneath
- Reddish thighs
- Obvious face-mask
- A dynamic, dashing sky-borne raptor

BACKLIT

- Strong light can have the effect of making underside look much darker

- Tends to look scythe shaped - often strongly recalls shape of swift in flight
- Slimline falcon that glides effortlessly, often very high

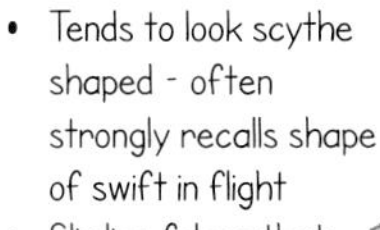

- Strong 'capped' effect
- Dark streaks down body - often make breast look simply dark when bird is high overhead
- Conspicuous, contrasting white throat
- Very obvious black moustachial stripe

- Dashing, dynamic hunter; interweaves soaring with sensational swoops and dives after prey
- Able to catch insects on wing and transfer them to bill in flight, so often seen lifting leg towards bill, something the other falcons do not do

Kestrel

SILHOUETTE

- Not a dramatic or fast flier - rather 'flappy'
- Longish tail
- Undernourished look to body
- Relatively long arm and hand

FEMALE

- Rich chestnut colour on upperparts
- Outer half of wing contrastingly darker than inner wing (useful feature)
- Lacks dark cap of Peregrine and Hobby

FEMALE

- Narrow but very well-defined black moustachial streak (c.f Merlin)
- Brown crown flecked with faint streaks

FEMALE, HOVERING

- Hovers (c.f. others to any extent)
- Although well barred, underwing light and sometimes glints brightly in sun

FEMALE

- In pursuit flight, wings flexed backwards and flies with fast, deep, stiff wingbeats
- Long tail always distinctive
- Tail pattern of regular dark bars on light brown colour, with broad dark band near end
- Warm upperpart coloration unique

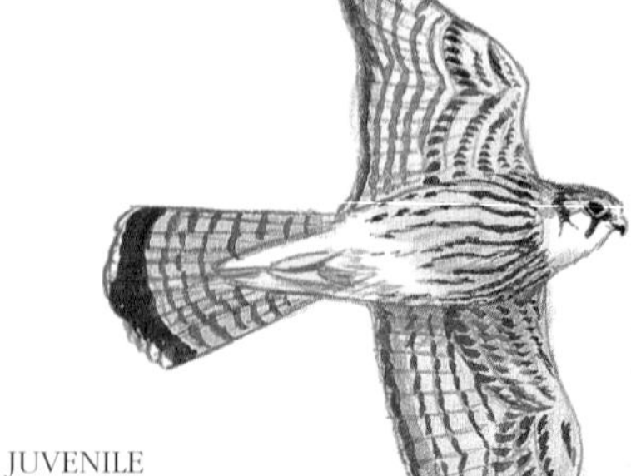

JUVENILE

- Poorly marked face-mask
- Wide black band on tail end
- Long-winged, long-tailed, weak-looking raptor

Merlin

SILHOUETTE
- Distinctly smaller than other falcons, often with rather light, flickering wingbeats, faster than those of the others
- Broad-based wings might recall Peregrine if size not obvious
- Medium-length, narrow tail - definitely shorter than Kestrel's, and square ended
- Hardly ever soars - usually moves along fast

CHASING SMALL BIRD
- Persistent 'guided-missile' pursuit of small birds diagnostic

MALE
- Deep buff almost to orange underparts (note: similar to male Sparrowhawk)
- Neatly and rather delicately streaked below ('dotted line', as Kestrel)
- Tail grey with broad dark band at tip
- Not very obvious moustachial streak

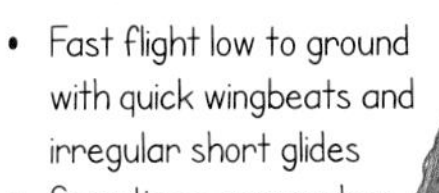

- Fast flight low to ground with quick wingbeats and irregular short glides
- Sometimes approaches prey using undulating flight
- Views usually fleeting
- Sharp-pointed, swept-back wings; when approaching prey, wings almost held closed

FEMALE
- Unlike male, white below with heavy dark streaking (fishbone-shaped streaks near flanks)
- Plain dark brown above (not grey)
- Larger than male (may approach Kestrel in size)
- Ladder-like bars on dark brown tail
- Like male, has poorly defined moustache and indistinct head pattern

JUVENILE
- Least developed face-mask
- Small, sharp-winged ferocious raptor (smallest European bird of prey, male being only slightly larger than Mistle Thrush)

RINGED, LITTLE RINGED & KENTISH PLOVERS

INTRODUCTION Common Ringed Plover (*Charadrius hiaticula*) [L 18cm] widespread, breeding on shingle beaches, riverbanks and tundra; in winter coastal, estuaries. Little Ringed Plover (*Charadrius dubius*) [L 16.5cm] summer visitor March–September; rather localized on gravelly margins of lakes, rivers and gravel pits. Widespread, although only in south in Britain and Scandinavia. Kentish Plover (*Charadrius alexandrinus*) [L 16cm] uncommon summer migrant April–October, to French and North Sea beaches, east to the west edges of the Baltic, on sandy places by the sea. Rare in Britain.

Plovers feed by alternating standing stock still, sentry-like, with rapid runs on fast-moving legs. Stop-start action distinguishes them from other waders.

Silhouettes

Ringed Plover

- Pointed rear end, but not as tapered as in Little Ringed; more balanced profile than in Kentish
- Short, fairly stout, blob-tipped bill
- Full chested
- Legs placed in middle of body

Little Ringed Plover

- Small, dainty, elegant and slim compared with the others
- May look more 'horizontal' than the others, and attenuated at rear
- Thin bill
- Rather small head (especially compared with Ringed)

Kentish Plover

- Ball-like shape, accentuated by strikingly short rear end = 'chick' shape
- Longer legs seem to be placed near back of body
- Finer bill than in Ringed

Head patterns

Ringed Plover

ADULT MALE
- Blunt, black-tipped orange bill
- Lots of black on head (c.f. Kentish)

JUVENILE
- Dark brown markings on head (black in adult breeding, but as adult winter)
- Dark ear-coverts have rounded lower border against white of collar
- Breast-band usually quite broad, especially at sides; sometimes broken or pinched in, as here (c.f. adult)
- Broad white supercilium diagnostic at this age

Little Ringed Plover

ADULT MALE
- Yellow eye-ring distinctive and diagnostic
- Slim, largely black bill, often with dull yellowish base
- White from supercilium extends up behind black on forecrown

JUVENILE
- Pale yellow eye-ring usually visible, but often not very clear
- Dark ear-coverts have pointed lower rear border against white of collar
- Buff supercilium, giving little contrast to surrounding head colour - may look 'hooded'
- Dull base to lower mandible of thin, black bill

Kentish Plover

MALE SPRING
- Chestnut coloration on crown variable, but usually present - different colour from back and mantle
- Long white supercilium meets white forehead
- Black bill
- Incomplete neck ring

JUVENILE
- Lots of white on forehead
- Collar interrupted, as in adult
- Breast patch only slightly darker than mantle (c.f. Ringed, Little Ringed)

Juveniles

Ringed Plover

- Scaly plumage on upperparts caused by feathers having white fringes
- Overall colour richer than in Little Ringed - said to be colour of wet sand, as opposed to dry sand (also applies to adults)
- Legs more colourful than in Little Ringed, often with orange tint, or at least yellowish
- Primaries peep out underneath tertials - fairly long primary projection

Little Ringed Plover

- Paler overall than Ringed juvenile
- Dull-coloured pink legs
- Primaries virtually covered by tertials - negligible primary projection

Kentish Plover

- Pale 'bleached' brown above; paler than the others
- Very white below - gleaming 'washing powder' white (similar to Sanderling)
- Subtle white fringes to feathers on upperparts
- Dark legs (black in adults)
- Collar incomplete - just mark on shoulder

Adults in flight

Ringed Plover

- Strong white wing-bar (as Kentish, but tail of Kentish very pale with broad white sides)
- Flight action fast and strong

Little Ringed Plover

- Very narrow, weak white wing-bar not visible in flight, instantly separating bird from other two species
- Longer winged and skinnier than Ringed
- Flight action whirring and fast, quite similar to a swallow or martin's

LARGE PLOVERS

INTRODUCTION European Golden Plover (*Pluvialis apricaria*) [L 27cm] breeds from Britain north to Arctic Scandinavia (also Iceland) on moorland and tundra; winters widely on salt marshes and pastures. Grey Plover (*Pluvialis squatarola*) [L 28cm] High Arctic nester outside region; winters on temperate muddy estuaries and beaches. Eurasian Dotterel (*Charadrius morinellus*) [L 22cm] summer visitor May–September, to mountaintops and stony tundra in Scandinavia, northern Britain and locally in the Netherlands (fields). On migration on agricultural fields, mountaintops, and so on.

JUVENILES

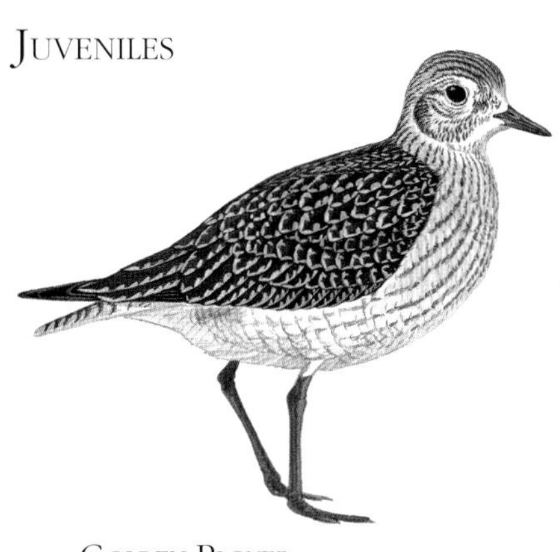

GOLDEN PLOVER

- Much thinner bill than Grey's
- Fine, pencil-thin streaks down breast denser than in Grey
- Spots on tertials, coverts and scapulars mainly golden

GREY PLOVER

- Juvenile confusingly golden tinged; easy to confuse with Golden
- Longer legs than Golden's
- Larger head than Golden's
- Spots on tertials, coverts and scapulars mainly whitish compared with Golden's, but note variation and use overall shape to identify

DOTTEREL

- Obvious white breast-band
- Warm buff below (c.f. adult)
- Plain tertials with pale edges (c.f. Golden, Grey)
- Scapulars mixture of young and old - may look rather messy

General ID tips

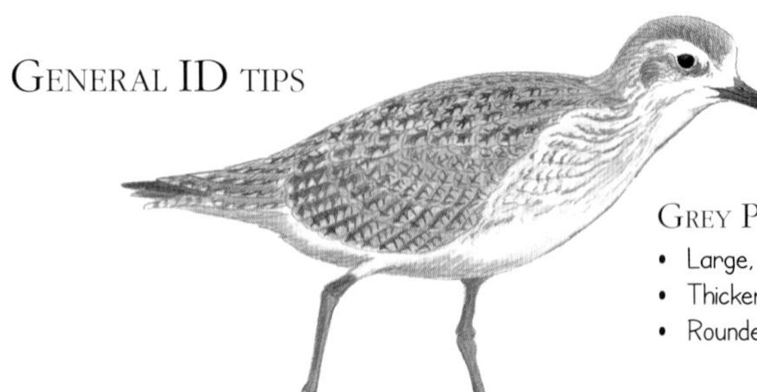

Grey Plover (adult winter)

- Large, heavy-looking wader
- Thicker bill than most waders
- Rounded head

Dotterel (adult winter)

- Slightly smaller than Golden Plover (much smaller than Grey)
- Greenish legs (c.f. Golden, Grey)
- Lacks spots of other two species: back scaly
- White breast-band

Dotterel (juvenile)

- Very broad supercilium continues to back of head, where it meets opposite supercilium to make 'V'
- Supercilium white in front; becomes buff behind

Golden Plover (adult winter)

Note: when disturbed, flocks have the habit of wheeling around in wide circles many times before landing again. Flocks oval in shape, or with 'V'-shaped leading edge - sheer height at which they often fly can be a useful pointer

- White underwing, which often shimmers white when flying overhead
- Dumpy body and very sharp wing-tips
- Narrow wings with very sharp wing-tips - noticeable at distance

Grey Plover (juvenile)

Note: these features hold good for all non-breeding plumages in flight

- Black 'armpit' (axillaries) diagnostic (and unique among waders generally)
- Broader, more obvious white wing-bar than Golden's
- Whitish rump

DUNLIN

Dunlin (*Calidris alpina*) [L 19cm] breeds on moorland, coastal grassland, salt marshes and tundra in Britain, Scandinavia and the Baltic region. Winters principally on estuaries, but also on inland lakeshores, and other wetland habitats. Often abundant, and the yardstick by which all other small waders should be compared. Universal Dunlin features include a black, slightly decurved bill; black legs; a little round shouldered; narrow white wing-bar; white rump split by black line down the middle; distinctive and bold in breeding plumage, but becomes grey and featureless in winter.

DUNLIN RACES IN BREEDING PLUMAGE (APRIL)

ALPINA

Breeds in northern Scandinavia and west Siberia. Winters in western Europe

- Largest and densest black belly patch of three races
- White band between belly patch and streaks on breast
- Richest chestnut on mantle
- A little more contrasting, with stronger white colour, than other races

ARCTICA

Breeds in Greenland, and passes through the area in late spring and autumn on its way to its West African winter quarters. Uncommon.

- Small black breast-patch
- Fine, speckled breast-spots
- Narrow grey-and-rufous fringes to mantle and scapular feathers

SCHINZII

Breeds in the Britain, Ireland, Greenland, Iceland and the Baltic region. Winters in West Africa.

- Breast-patch intermediate in size between other races
- Spots on breast meet or merge with black patch
- Some mantle and scapular feathers have definite yellowish tinge to fringes

DUNLIN BILL LENGTH

JVENILE,
ONGER BILLED

Length of bill indicates *alpina* or *schinzii*

Degree of curvature varies

JUVENILE,
SHORTER BILLED

- Typical *arctica* bill, probably male's

Dunlin (selection of bills)

- Note: length and curvature of Dunlin bills varies greatly, according to sex (males have shorter bills on average than females, by 3-5mm) and race. It is thus hard to identify races just by the length of their bills, especially in the field.

Longest and more curved bill - often *alpina*

Shortest bill - often *arctica*

Ageing Dunlin

juvenile, Early August

- Most distinctive feature (almost unique) black dots on breast and belly, which often form dotted lines - degree of spotting varies, but is always present
- Distinctly rufous-toned head and breast
- Buff crown
- Strong streaking on breast

juvenile, late August/early September

- Warm brown head and neck, with fine streaking, especially on crown
- Some wing-coverts and scapulars dark with pale or warm brown fringes, typical of full juvenile plumage; in this plumage often shows silvery 'V' on scapulars
- Some wing-coverts grey with pale fringes, typical of first-winter plumage

juvenile, flying

- Narrow but always visible white wing-bar
- Black bar splits white rump

earlier juvenile–1st winter, mid-September

- Spotting on upper breast betrays this is juvenile bird
- Most of copious spotting has disappeared, but has not all been 'rubbed out'
- Grey 1st-winter feathers start to appear on mantle

JUVENILE-1ST WINTER, MID-SEPTEMBER
Shows neat mixture of:

- Juvenile plumage - remnant of black dots, brown on crown and nape, a few dark wing-coverts
- (First) winter plumage - mainly greyish upperparts, mainly whitish underparts

JUVENILE-1ST WINTER, LATE SEPTEMBER

- Spotting on breast variable - diminishes as season progresses, but also varies between individuals, with some much better marked than others
- Grey patches on plumage indicate progression to winter plumage

1ST WINTER, OCTOBER

- First-winter and adult-winter plumage almost identical, except for small differences in fringes of wing-coverts and tertials
- Mainly grey-brown above
- Inconspicuous but well-defined breast-band formed by contrast between white belly and greyish streaks
- Not very obvious whitish supercilium on greyish head

LATE 1ST WINTER

- Compared to summer, very featureless and dull (good distinguishing trait from many other species)
- Very faint streaking on breast
- Pale grey breast-sides with sharp contrast to clean white belly

VINTER, FLYING

- Rather grey and featureless
- Narrow white wing-bar in dark upperwing
- Rump white except for bold black stripe down middle, which splits it in two

DUNLIN & CURLEW SANDPIPER

INTRODUCTION Curlew Sandpiper (*Calidris ferruginea*) [L 21.5cm] breeds central Siberia, and passes through Western Europe July–September; less common in spring. Mainly coastal lagoons and mudflats. (For Dunlin introduction, see page 97-99.)

Dunlin (juvenile)

- Not very even rows of spots down breast and centre of belly (unique)
- Upper breast richer brown colour than in Curlew Sandpiper
- Reddish-brown colour on head, darker than in Curlew Sandpiper
- Birds in fresh plumage (earlier in season, July-August) show one or two narrow white stripes along 'shoulder', lacking in Curlew Sandpiper (similar to juvenile Little Stint)
- In contrast to Curlew Sandpiper, feather fringes on scapulars often warm rust-red

Curlew Sandpiper (juvenile, July-late Sept)

- Distinctly longer legs than Dunlin's - often paddles in deeper water
- Paler and more pastel coloured than juvenile Dunlin
- Upperparts densely, evenly and neatly scaled; coverts have pale tips, inside which are dark 'anchor' marks, very different from Dunlin

Curlew Sandpiper (juvenile)

- Much neater and more cleanly marked than juvenile Dunlin
- Always has longer and more evenly curved bill than Dunlin's
- Well-marked pale supercilium

Dunlin (1st winter, October)

- Adults acquire winter plumage from early autumn (August) onwards and can look like this when mixing with Curlew Sandpipers - note similarity to juvenile Curlew Sandpiper
- Inconspicuous but well-defined breast-band formed by contrast between white belly and greyish streaks; this can invite confusion with juvenile Curlew Sandpiper, but this has peachy wash to breast
- Not very obvious whitish supercilium on greyish head

Comparison in flight

Dunlin (winter)

- Greyish and featureless; much cleaner underparts than in juvenile Dunlin, hence can be confused with Curlew Sandpiper
- White wing-bar much the same as in Curlew Sandpiper
- Smaller and slightly more compact than Curlew Sandpiper

Dunlin (juvenile)

- Black centre to rump and tail
- Tail with grey sides

Curlew Sandpiper (juvenile)

- In among flock of Dunlins, larger size may be apparent
- Buff wash to upper breast
- White rump key feature

Dunlin (flying away)

- Clear white wing-bar
- Dark stripe splits rump and tail
- Flies with very rapid wingbeats, usually very low and often in tight, well-coordinated flocks

Curlew Sandpiper (flying away)

- Clear white wing-bar, as in Dunlin
- White rump obvious
- Legs project beyond tail

TEMMINCK'S & LITTLE STINTS

INTRODUCTION Temminck's Stint (*Calidris temminckii*) [L 14.5cm] summer visitor to Scandinavia May–September, breeding in short grass and open areas near water; sometimes near human cultivation. Passage migrant to rest of area, in freshwater marshes and pools. Little Stint (*Calidris minuta*) [L 14.5cm] breeds on tundra in the high Arctic. It is a common passage migrant, especially in autumn, to saline and fresh pools and lagoons.

Little Stint has feverish feeding style. It is more sociable than Temminck's, which is often solitary. The adage 'Temminck's looks like miniature Common Sandpiper' almost always works. It feeds quite slowly and tends to crouch on flexed legs and 'creep' along. Little is quite tame and does not flush far; Temminck's flies off very fast and high.

General impression

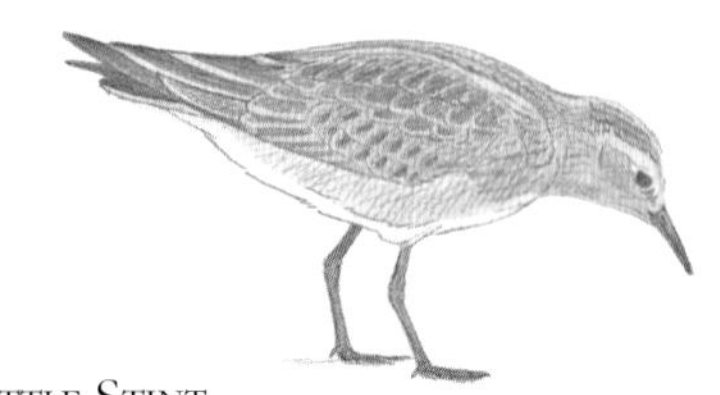

Little Stint

- Plumper than Temminck's
- Often leans down to feed
- Wings stick out beyond tail (c.f. Temminck's)

Temminck's Stint

- Paler legs than Little's (green, not black)
- Tail projects beyond long wings, giving tapered rear end
- Shorter legs than Little's

Ageing Temminck's Stint

Adult spring

- In this plumage often with quite warm coloration
- Rusty edges to some coverts and scapulars (varies between individuals)
- Black-centred scapulars
- White outer-tail feathers (grey on Little) - hard to see

Adult summer

- Note typical horizontal, low-slung carriage
- Soberly coloured plumage without rusty tones
- Green legs
- Summer adults adorned with black-centred scapulars

Juvenile

- Unlike Little, looks similar to its equivalent adult; colourless compared with Little
- Plain head without obvious whitish supercilium
- Obvious breast-band, as adult
- Upperparts scaly, with coverts and scapulars having narrow buff fringes and blac subterminal fringes

Ageing Little Stint

ADULT, SPRING MOULT

- Adults variable in spring, and some may show combination of grey-brown winter plumage and colourful summer plumage

ADULT SUMMER

- Much brighter coloured than dowdy Temminck's
- Warm chestnut colour on edges of coverts
- Pale supercilium 'splits' in front of eye
- Streaks may be confined to sides of breast

JUVENILE

- White 'V'-mark on edge of mantle very obvious - a lot more so than in adult; often a second 'V' on scapulars
- Immaculate white belly (lacks spots of Dunlin juvenile)
- Streaks with buff background mainly on breast-sides
- Many coverts blackish in middle, with pale fringes

JUVENILE

- 'Split supercilium' - with one branch over eye and one below crown
- Dark centre to crown

UVENILE MOULTING TO 1ST WINTER

Colourful juvenile fades quickly to look like this by October

White 'V' scarcely visible

Retains many dark juvenile coverts

ADULT WINTER

- In winter, dull-coloured wader without much contrast
- Thin, straight, spiky bill, proportionally shorter than Dunlin's
- More white on neck than in Dunlin
- Black legs (as Dunlin

SANDERLING

INTRODUCTION Sanderling (*Calidris alba*) [L 19cm] is a high Arctic nester and reasonably common passage migrant and winter visitor, mainly on sandy beaches around Britain and west-facing coasts from Denmark southwards.

Ageing Sanderling

ADULT WINTER

- 'Washing-powder white' underparts
- Diagnostic dark patch at base of wing ('shoulder') - sometimes hidden by white plumage around it
- Black legs
- Sturdy, shortish black bill

ADULT WINTER, RUNNING

- Often seen as very white wader running along beach like a clockwork toy
- Very pale grey above; overall much paler than Dunlin (and stints)
- Typically feeds in small flocks on sandy beaches, usually at tideline, dashing back and forth as the waves break, and picking rapidly for small creatures exposed by waves

ADULT WINTER, FLYING

- Powerful flight with whirring wingbeats
- Broad white wing-bar (more conspicuous than in any other numerous small wader)
- Wing otherwise mainly blackish
- Body looks very white

ADULT, APRIL

- Mix of summer and winter plumages, with pale grey of winter being replaced by chestnut tones of summer on upperparts
- Newly grown coverts rufous with black markings and pale fringes
- Beginnings of breast-band

DULL ADULT, APRIL

- Much duller bird than average; probably youngster gaining first-summer plumage
- Appearance variable, but mixture of juvenile and breeding plumage
- Rusty colour on some coverts

ADULT, MAY

- Darker plumage than in spring; a more intense rufous colour
- Whole of head and neck unremittingly reddish-brown, with black peppering
- Very strongly defined breast-band

ADULT, AUGUST

- Variable clash of summer and winter plumages; pale grey and rufous
- Breast-band still shows
- White belly still useful guide

JUVENILE

- Very distinctive black-and-white-chequered pattern on back and mantle
- Dark crown and broad white supercilium
- Very clean white below
- Limited streaks on breast sides

JUVENILE, RUNNING

- Distinctive plumage, but running habit remains the same

JUVENILE–1ST WINTER

- Pale grey above, as adult
- Retains some juvenile plumage on coverts and tertials (amount reduces towards end of year, and looks like adult from about February)

RUFF

INTRODUCTION As a breeding bird Ruff (*Philomachus pugnax*) [L 31cm male, 24cm female] is locally common in Scandinavia, the Baltic region and the Low Countries, on meadows, marshes and bogs. It is a fairly common migrant throughout the region and winters sparingly from Britain southwards – but mostly in Africa.

Males are much larger and bulkier than females; the distinction usually quoted is that a male can be as large as a Redshank and the female as small as a Dunlin. Despite their variation, all Ruffs have distinctly small heads that look too small for their bodies. The body is rather bulky and dumpy, the neck long and the bill of medium length and slightly downcurved. A Ruff walks with an odd gait, with the head moving in step like a pigeon's.

Ageing and sexing Ruff

MALE SUMMER, BLACK
- Individual Ruffs with black plumage often dominant and remain at same breeding ground (lek) through whole breeding season

MALE BREEDING (LATE APRIL–JUNE)
- Male with brown 'ruff' - totally unmistakable, but head and neck plumes are lost quickly after breeding

MALE BREEDING
- Every male exhibits slightly different finery, be it in pattern or colour, making birds individually recognizable

MALE BREEDING, SATELLITE
- White 'ruff'

MALE SPRING (JANUARY-APRIL)
• Remarkably, this is a male in breeding plumage that has yet to grow its colourful ruffs. These grow late April-early May, with bare skin on face appearing a little later
FEMALE BREEDING
• Hard to conceive this dowdy bird is same species as breeding male
• Usually strongly grey-brown head and upper breast
• Scaly back with pale tips to coverts, and often quite warm in coloration
FEMALE BREEDING
• May have quite dark and dense head pattern
MALE MOULTING (JULY)
• Male finery lost remarkably early, first disappearing in June
MALE MOULTING (JULY)
• Black blotching on breast typical
• Whitish head typical (may remain throughout winter)
• Mix of breeding and new winter plumage on back

ADULT WINTER FEMALE

- Back plumage typically ruffled and loose - can be useful ID feature
- Scaly pattern a good feature, and much more obvious than in most other small to medium-sized waders - however, still not as scaly as juvenile
- Reddish legs point to adult bird (juveniles and first winters have greenish legs)
- White bill-base, forming 'sheepskin nose-band'

ADULT WINTER MALE

- Larger and bulkier than female
- Males have relatively shorter bills than females
- White eye-ring
- Orange-red legs (typical of adult, but colour varies in intensity)
- Red bill-base typical of adult

ADULT WINTER MALE, ALTERNATE

- Notable proportion of males retain some white on head and underparts during winter, making them more distinctive than typical birds

JUVENILE

- Warm plain buff colour on head and underparts
- Juveniles vary in colour - this bird is less intensely buff than some
- Legs dull green to brownish (orange in adults)
- Pleasingly neat, scaly pattern on wings and back

JUVENILE FEMALE
• Sexual size differences apply as for adults - this can be very confusing in a flock, especially when other species are around
• Smallest females only a little larger than Dunlin
JUVENILE MALE
• Larger (Common Redshank size) and podgier than female
JUVENILE
• Medium-length bill
• Distinctive oval-shaped white patches either side of rump (very useful feature in flight)
• Narrow white wing-bar (widest on inner primaries)
• Short tail with legs sticking out behind
FLIGHT SILHOUETTE
• Feet stick out beyond tail
• Often glides when circling or about to land (distinctive)
• Wings do not beat at same super-fast rate as many waders' - more like pigeon pace
1ST WINTER
• Narrow white wing-bar
• Looks very grey-brown
• Legs greenish

Some Ruff lookalikes

Ruff (juvenile)

- Medium-length bill
- Distinctive oval-shaped white patches either side of rump
- Narrow white wing-bar
- Legs project beyond tail

Buff-breasted Sandpiper (juvenile)

- Dark rump without white ovals
- Legs do not obviously trail behind
- No wing-bar above
- Shorter, straighter bill than Ruff's; all black
- Spots on sides of breast
- Yellowish legs
- Dark, pale-rimmed eye
- Peppered spots on crown

Curlew Sandpiper (juvenile)

- White rump
- Much faster wingbeats than Ruff's

Ruff (juvenile)

- Warm plain buff colour on head and underparts
- Juveniles vary in colour - this bird is less intensely buff than some
- Legs dull green to brownish
- Neat, scaly pattern on wings and back

Curlew Sandpiper (juvenile)

- Prominent supercilium
- Black legs
- Bill much longer and more decurved than Ruff's
- Buff colour on underparts limited to breast

Pectoral Sandpiper (juvenile autumn)
- Overall colour and pattern quite similar to juvenile Ruff's
- Breast streaks (c.f. juvenile Ruff)
- Narrow white stripes or 'braces' on back

Pectoral Sandpiper (adult autumn)
- Slightly smaller than Ruff
- Pectoral band = breast/belly contrast
- Greenish legs
- Dark eye-stripe

Pectoral Sandpiper (juvenile)
- Sharp contrast between white belly and breast streaks against buff background

Dunlin (adult)
- Black, downcurved bill
- Dark eye-stripe
- Legs much shorter than Ruff's

Ruff (winter)
- Bill curves down
- White at bill-base
- Paler grey overall than Common Redshank

Common Redshank (adult winter)
- Straighter bill than winter Ruff's
- Legs may be similar colour to Ruff's
- Plain back (no scaling)

Common Redshank
- Prominent white trailing edge to wings instantly identifies it

COMMON, GREEN & WOOD SANDPIPERS

INTRODUCTION Small freshwater waders that have the unusual habit of persistently wagging their rear ends up and down. Common Sandpiper (*Actitis hypoleucos*) [L 19cm] common and widespread summer visitor April–September, and passage migrant beside rivers, lakes and pools, especially in uplands. Some birds overwinter. Green Sandpiper (*Tringa ochropus*) [L 22cm] summer visitor to damp northern forests in Scandinavia and around the Baltic; some birds winter in Britain, France and the Low Countries, beside streams and lakes. Wood Sandpiper (*Tringa glareola*) [L 20cm] summer visitor May–September, breeding in well-wooded marshes and bogs in Scandinavia; passage migrant elsewhere, mainly to fresh water.

COMMON SANDPIPER (ADULT SUMMER)
- Warm olive-brown upperparts contrast sharply with clean white underparts
- Sharp demarcation between brownish breast and white belly
- Characteristic white 'comma' at wing edge
- Horizontal carriage, but habit of incessantly wagging rear end makes it look as though it leans forwards

COMMON SANDPIPER (JUVENILE)
- Smaller and slimmer than Green and Wood, with shorter legs
- Straight, fairly broad-based bill about as long as head
- Tail sticks out beyond wings (c.f. others)
- Pale tips to coverts and scapulars give lightly scaled effect (c.f. adult)

GREEN SANDPIPER (JUVENILE)
- Overall gives strongly black-and-white appearance - very dark above and clean white below
- Dividing line between upperparts and underparts runs more or less level from undertail-coverts to sides of breast
- Clear breast-band (c.f. Wood)
- Strongly marked supercilium in front of eye only (see Wood Sandpiper)

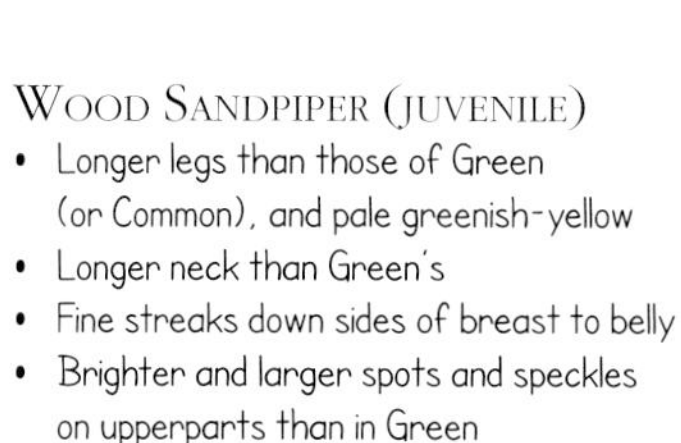

Wood Sandpiper (juvenile)

- Longer legs than those of Green (or Common), and pale greenish-yellow
- Longer neck than Green's
- Fine streaks down sides of breast to belly
- Brighter and larger spots and speckles on upperparts than in Green
- Much more obvious supercilium behind eye than in Green

Common Redshank (juvenile)

- Slightly larger than Wood, with longer legs
- Bill-base orange-pink
- Orange legs
- Heavier streaks on breast and belly than in Wood

Juveniles in flight

Wood Sandpiper

- Narrower wings than Green's, especially at base
- Close barring on tail
- White rump
- Legs stick out well beyond tail

Green Sandpiper

- Often flushed from ditches and streamsides - flies off with rapid, zigzagging flight
- Bright white rump against blackish upperparts may bring to mind House Martin
- Dark underwings contrast sharply with white flanks and belly
- A few bold black stripes on tail

Common Sandpiper

- Flight pattern unique and distinctive: flies very low over water with very fast, shallow and often intermittent wingbeats interspersed with short glides
- Distinctive bold black markings on underwing
- Upperwing has broad wing-bar all along it (not in Green or Wood)
- Tail brown with barred edges

COMMON REDSHANK, GREENSHANK & SPOTTED REDSHANK

INTRODUCTION Common Redshank (*Tringa totanus*) [L 25cm] widespread and common breeding bird in wet grassland, salt marsh and bog; chiefly winters on coastal salt marsh and muddy estuaries. Common Greenshank (*Tringa nebularia*) [L 32cm] breeds in northern Scotland and Scandinavia on taiga bogs; winters sparsely on mudflats in Britain southwards, but most birds go to Africa. Spotted Redshank (*Tringa erythropus*) [L 31cm] breeds in northern Scandinavia in forest tundra, wintering mainly south of the region; fairly common passage migrant in between.

Commonly referred to collectively as 'shanks', these are medium-sized, long-legged and long-billed waders. Their bills are essentially straight and their bodies slim and elegant.

WINTER ADULTS

COMMON REDSHANK

- Orange-red legs
- Bill has orange base to both mandibles, and black tip
- Much browner plumage than Spotted Redshank's
- Shorter, entirely straight bill compared with that in Spotted

SPOTTED REDSHANK

- Black eye-stripe contrasts strongly with white lores and supercilium
- Bright red legs (look brighter than Common's)
- Whiter on underparts than Common
- Long bill has bright red only at base of lower mandible, contrasting with black colour of rest of bill
- Bill has slight droop at tip (may look as though water drop is on end of it)

GREENSHANK

- At distance looks whitish
- Long, slightly uptilted, dark grey bill
- Very white below, with strong contrast to upperside
- Larger than Common or Spotted Redshank

Shanks' in flight

Common Redshank (adult winter)

- White trailing edge
- Outer wing darker than inner wing
- Many dark bars on tail (c.f. Greenshank)
- Slightly broader white rump and back than in Spotted
- Shorter legs than in Spotted

Greenshank (juvenile)

- Plain dark wings (c.f. Common Redshank)
- Prominent white rump, making 'V' well up back
- A few diffuse bars on tail, but tail partly white (c.f. Common and Spotted Redshanks)

Spotted Redshank (juvenile)

- Note: flight features hold good for all non-breeding plumages.
- Whole upper wing looks dark when bird flies; white closed oval mark on back
- Tail white with black bars (looks greyish)
- Legs project well beyond tail (see Common)

Juveniles

Common Redshank

- Bill slightly longer than head
- Bill straight
- Dull orange bill-base (c.f. adult)

Greenshank

- Longer bill than Common Redshank's, and blue-grey
- Much heavier and thicker bill than the others, with broad base
- Bill uptilted
- Less patterned head than in the others

Spotted Redshank

- Very long bill with red on lower mandible (as adult)
- Very prominent white loral patch above eye-stripe

Greenshank

- Strong contrast between dark upperparts and whitish underparts and head - typical of species
- Greenish legs
- Darker above than other juveniles, and with scaly pattern rather than spots
- More streaked on head and neck than adult

Common Redshank

- Legs paler than adult's
- More streaked on underparts than adult (Spotted juvenile is barred, Greenshank juvenile less streaked)
- Coverts and scapulars have buff fringes (not adult)

Spotted Redshank

- Dusky greyish-brown wash
- Strongly barred on underparts, from neck to undertail-coverts
- Upperparts with many small white spots
- Legs not as red as adult's

Ageing Spotted Redshank

JUVENILE

- See page 116 for features
- Inveterate swimmer at any age (good identification pointer)

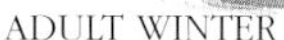

ADULT WINTER

- Pale grey above (lacks brown tones)
- Prominent white supercilium (mainly before eye, as in all non-breeding plumages)
- Red legs make stark contrast to dull plumage
- Slightly drooped tip to bill

1ST WINTER

- A few bars remain on underparts into October
- Upperparts with darker ground colour than adult's

ADULT, APRIL

- Remarkable sooty colour of full breeding bird acquired March-May - this bird is in fresh plumage (note pale feather edges)

ADULT, MAY

- In breeding plumage, handsome and looks like no other wader
- Clean coal-black colour
- Small white dots on wings
- Legs black

ADULT, JULY

- Distinctive breeding plumage soon lost - this bird is effectively in winter plumage with a few dark blotches

GODWITS

INTRODUCTION Godwits are large waders with long legs and long, straight bills, often with orange on the plumage. Oddly quiet for waders. Black-tailed Godwit (*Limosa limosa*) [L 41cm] breeds in continental Europe and Iceland on grassy meadows and damp pasture; winters locally on coastal marshes and mudflats. Bar-tailed Godwit (*Limosa lapponica*) [L 39cm] breeds in Arctic Russia; winters on widely on estuaries and beaches.

Black-tailed Godwit Races

limosa Breeds in continental Europe, including eastern England; winters in Iberia, North Africa and Africa south of the Sahara.

islandica Breeds in Iceland; winters from Britain southwards.

Winter godwits

Bar-tailed Godwit

- Upperparts streaked, rather like a Curlew's
- When identifying tricky godwits, check 'knees' (actually tibiotarsal joint, equivalent to our ankle). In Black-tailed, distance between 'knee' and belly is long (enough to write word 'BLACK'); in Bar-tailed, shorter and only enough to write word 'BAR'

- Supercilium projects further behind eye than in Black-tailed
- Bill tends to look slightly more uptilted than Black-tailed's

Black-tailed Godwit

- Plain grey-brown upperparts, without streaks
- Very upright and elegant
- Walks smoothly with grace, without slight awkwardness of Bar-tailed
- A confirmed paddler, often feeding in deep water up to belly; habitually submerges head underwater

- Plain head without streaks
- Supercilium reaches eye but not far beyond

BLACK-TAILED GODWIT
- Huge white wing-bar impossible to miss
- Black tail and square-shaped white rump
- Long legs project well beyond tail
- Dark-edged underwing

BAR-TAILED GODWIT
- Like 'Curlew with straight bill'
- Tail with close black bars (detail difficult to see)
- White rump, colour extending backwards to make 'V' (similar to Curlew's pattern)
- Shorter legs do not project far beyond tail

JUVENILE GODWITS

BLACK-TAILED GODWIT (*ISLANDICA*)
- Intense orange-buff coloration on throat and breast

BLACK-TAILED GODWIT (*LIMOSA*)
- Paler colour on throat and breast than in *islandica*
- Often slightly larger and taller than *islandica*, with deeper bill-base giving less rounded forehead

BLACK-TAILED GODWIT (*LIMOSA*, MOULTING TO 1ST WINTER)
- Odd combination of juvenile and adult-like winter plumage Body colour browner than adult's

BLACK-TAILED GODWIT (*LIMOSA*)
- Tertials with some grey coloration and even pale fringes
- Orange tone to neck and breast
- Always shows some obvious large, dark spots on wing-coverts
- Always shows some warm, rusty feather-tips

BAR-TAILED GODWIT
- Very different upperpart pattern from juvenile Black-tailed's: streaked and Curlew-like
- Tertials evenly marked
- Rather wan, buff colour to neck and breast
- Some streaking on belly and flanks (c.f. Black-tailed)

CURLEW & WHIMBREL

Eurasian Curlew (*Numenius arquata*) [L 52cm] breeds widely on moors, bogs, meadows, and so on; winters on estuaries, coasts and fields. Eurasian Whimbrel (*Numenius phaeopus*) [L 42cm] breeds in Scotland, Iceland and Scandinavia on moorland and bogs up to the tundra, where it is a summer visitor May–August. Widespread on passage on rocky shores, beaches, reefs and coastal grassland; winters mainly in West Africa.

Whimbrel

- Extremely long, downcurved bill, large size and long legs render Curlew and Whimbrel unmistakable
- Slightly darker in appearance than Curlew
- Slightly shorter neck and more angular head than Curlew's (subtle difference)

Curlew

- Bill longer and more evenly curved than Whimbrel's
- Straw-coloured plumage with intricate streaks typical of both species - very few plumage details differ significantly
- Much larger than Whimbrel

Head and bill

Whimbrel (partly facing)

- Striking pattern to crown - pale cream centre with brown lateral stripes either side; some Curlews may exhibit indistinct pale crown centre

Whimbrel (from side)

- Curve of bill exaggerated towards tip (inner half relatively straight), so looks kinked rather than evenly curved
- Shorter bill than in any Curlew

Curlew (juvenile male)

- Shortest Curlew bill, significantly shorter than female's bill and potential pitfall compared with Whimbrel
- Base of lower mandible often much paler than rest (c.f. Whimbrel)

Curlew (adult female)

- Longest bill - longer than in adult male and juvenile
- Featureless face compared with Whimbrel
- White eye-ring more prominent than in Whimbrel
- Streaked crown

In flight

Curlew

- Large white 'slit' up back (as Whimbrel)
- Lacks wing-bars - pattern more like immature gull's than that in most waders
- Slow wing-beats (gull-speed)
- More contrast between dark outer wing and paler inner wing than in Whimbrel
- Feet stick out slightly beyond tail

Whimbrel

- Faster wingbeats than those of Curlew, with distinctively stiffer action
- Overall slightly darker and less contrasting wings than in Curlew

SNIPES & WOODCOCK

INTRODUCTION Common Snipe (*Gallinago gallinago*) [L 27cm] widespread resident in marshes, wet meadows and floods; summer visitor to Scandinavia and Iceland. Eurasian Woodcock (*Scolopax rusticola*) [L 36cm] widespread resident in damp woodland; summer visitor to Scandinavia. Jack Snipe (*Lymnocryptes minimus*) [L 19cm] rare breeder in large bogs in Scandinavia, wintering widely in similar habitats to Common Snipe, but locally. Great Snipe (*Gallinago media*) [L 28cm] rare (extremely local breeder in Scandinavia), and included here as an occasional passage migrant; very rare in Britain.

Jack Snipe is easily picked out by its quirky habit of persistently rocking its front body up and down when feeding, as if on springs. Snipe and Woodcock do this sometimes, but not persistently or habitually.

COMMON SNIPE
- Dumpy, short-legged and with cryptic, streaky brown plumage
- Underparts look barred, not streaked, including on flanks
- White in tail (c.f. Jack Snipe)
- White belly (c.f. Great Snipe)

GREAT SNIPE
- Three broad, wavy white bars across mid-wing (showing as reasonably prominent bars in flight)
- Heavily barred across breast and flanks, right down to belly
- Much more white on tail sides than in Common
- Broader supercilium before eye than in Common

JACK SNIPE
- Looks dumpy and neck-less
- Two broad, very prominent golden stripes along back
- Dark mantle with green sheen
- Breast with limited streaking, and white belly

COMMON SNIPE
- Pale central crown stripe (c.f. Jack Snipe).
- Single pale supercilium below dark crown

JACK SNIPE
- Much shorter bill than Common's
- Bill looks broader at base than in Common
- Short black 'eyebrow' within broad creamy 'double' supercilium
- Ear-coverts with black surround

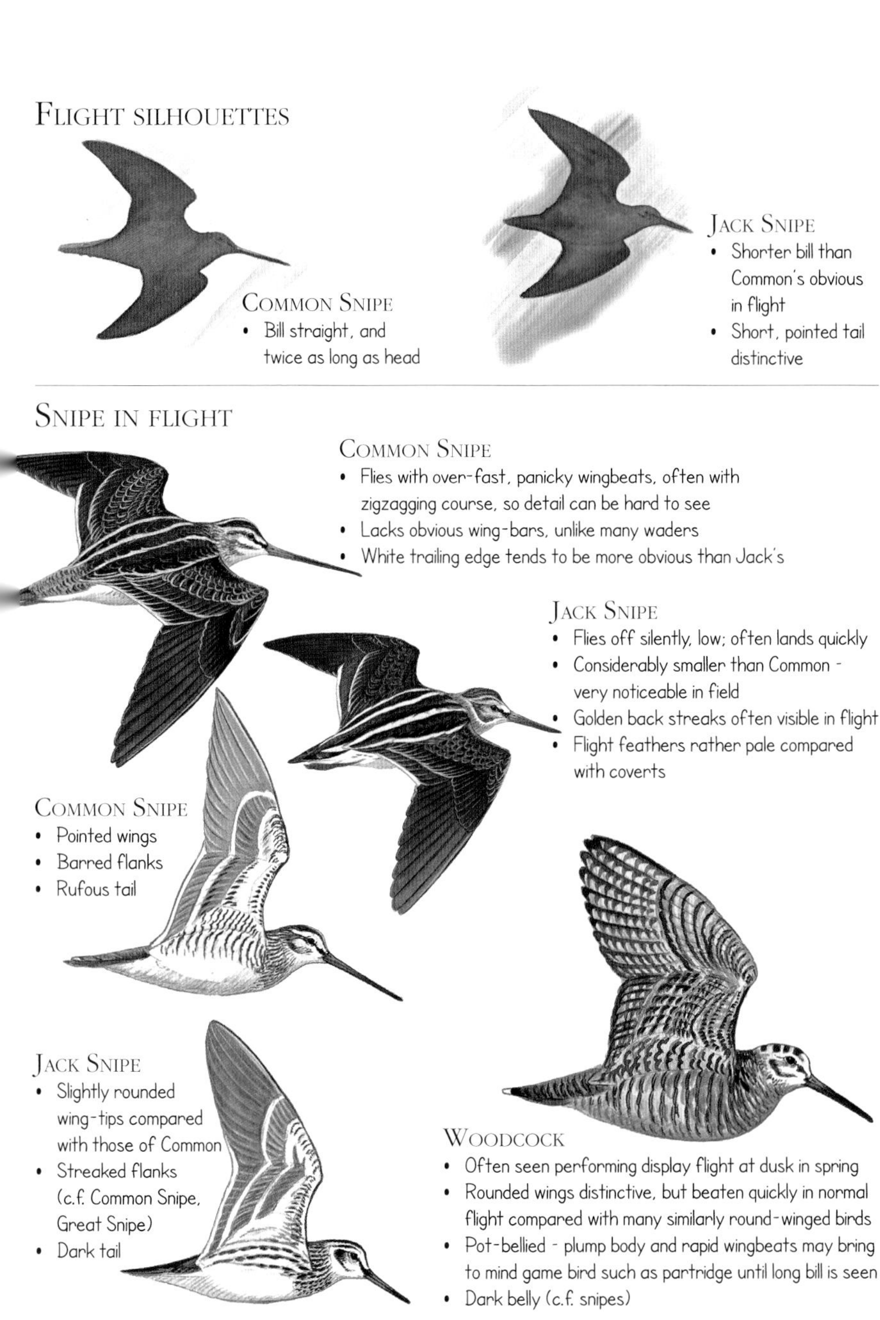
Flight silhouettes
Common Snipe
• Bill straight, and twice as long as head
Jack Snipe
• Shorter bill than Common's obvious in flight
• Short, pointed tail distinctive
Snipe in flight
Common Snipe
• Flies with over-fast, panicky wingbeats, often with zigzagging course, so detail can be hard to see
• Lacks obvious wing-bars, unlike many waders
• White trailing edge tends to be more obvious than Jack's
Jack Snipe
• Flies off silently, low; often lands quickly
• Considerably smaller than Common - very noticeable in field
• Golden back streaks often visible in flight
• Flight feathers rather pale compared with coverts
Common Snipe
• Pointed wings
• Barred flanks
• Rufous tail
Jack Snipe
• Slightly rounded wing-tips compared with those of Common
• Streaked flanks (c.f. Common Snipe, Great Snipe)
• Dark tail
Woodcock
• Often seen performing display flight at dusk in spring
• Rounded wings distinctive, but beaten quickly in normal flight compared with many similarly round-winged birds
• Pot-bellied - plump body and rapid wingbeats may bring to mind game bird such as partridge until long bill is seen
• Dark belly (c.f. snipes)

SKUAS

Great Skua (*Stercorarius skua*) [L 55cm, WS 139cm] breeds mainly on offshore islands in Scotland, Faeroes, Norway and Iceland, on moorland; winters at sea, mainly in the Atlantic. Arctic Skua (*Stercorarius parasiticus*) [L 40cm, WS 113cm] breeds on coasts and islands of Scotland, Scandinavia, Iceland and Faeroes; winters at sea mainly south of the region, but off most coasts on passage. Pomarine Skua (*Stercorarius pomarinus*) [L 46cm, WS 120cm] breeds in the Arctic outside the region; uncommon passage migrant on coasts. Long-tailed Skua (*Stercorarius longicaudus*) [L 37cm, WS 106cm] summer visitor to northern Scandinavia, on tundra; scarce passage migrant on coasts elsewhere.

Skuas are gull-like but have dark plumage and pointed wings that give them an almost falcon-like shape. They also have a far more purposeful, menacing flight than gulls. Several of the region's skua species occur in two colour morphs, dark and pale, with a few intermediates between the two. All Great Skuas look much alike.

Long-tailed is the size of Black-headed Gull, Arctic the size of Common Gull; Pomarine matches Lesser Black-backed Gull. Pomarine is sometimes mistaken for Great, an error that would never be made about an Arctic. Pomarine flies with steady, languid progress, while Arctic is more dashing. Great is the only species to chase Gannets, while Pomarine may go for large gulls, which are too much for Arctic.

Adults in spring

Great Skua

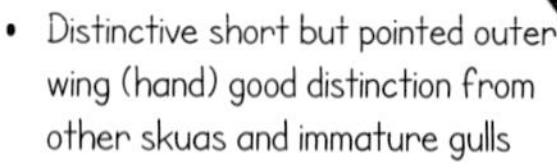

- Distinctive short but pointed outer wing (hand) good distinction from other skuas and immature gulls
- Very short, wedge-shaped tail (completely different from tails of other skuas; gulls have longer tails that are square ended)
- All over dark brown with straw-like pale streaks and blobs
- Wing-flashes extensive (more so than in other skuas; completely lacking in immature gulls)

Pomarine Skua (pale morph)

- Barrel chested, with larger bill and head than Arctic
- Tail-streamers 'spoon shaped' in side view (twisted 90 degrees to horizontal) - tail looks blunt even if spoons are lost
- Black on head reaches down to below bill-base
- Often larger wash of pale yellow on sides of neck than in Arctic
- Most (but not all) have mottled breast-band and barring on flanks

Arctic Skua (pale morph)

- Long, narrow wings, thinner than those of Pomarine and Great
- Much slimmer in chest and belly than Pomarine
- Tail-streamers pointed, not blunt, and extend out about half length of tail
- White on underparts often extends further down body than in Pomarine, almost down to undertail-coverts

Long-tailed Skua

- Long rear end and delicate build highly distinctive
- Very long tail-streamers; tail 'spike' flexes and quivers in wind
- Very little white on primaries - just streak above and below
- Inner wing colour (pale grey-brown) contrasts strongly with outer wing colour (very dark brown)
- Whitish colour on throat and belly gradually merges into darker grey-brown on belly and undertail-coverts (sharper distinction in Arctic and Pomarine)

2nd autumn

Arctic Skua

- Black cap (as adult)
- Heavily barred underneath
- Single wing-flash
- Hint of breast-band

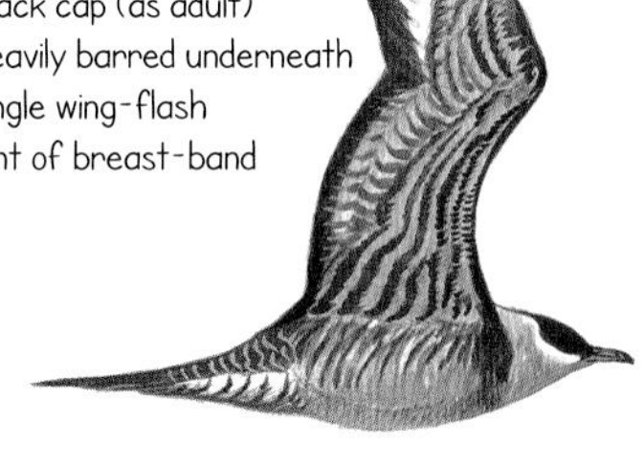

Pomarine Skua

- Barring on underside indicates immaturity
- Familiar head and belly pattern similar to adult Pomarine's
- Strong chest band
- Double wing-flash
- Stronger barring on undertail than in Arctic of similar age

Long-tailed Skua

- Many pale juveniles have distinctly pale (or 'creamy') heads like this
- Many have pale belly and dark chest band
- Very strongly barred on flanks, underwings and undertail-coverts

Juveniles

Great Skua

- Basically similar to adult (other immature skuas are less so), but darker
- Intense chocolate-brown below without barring or speckles (c.f. adult)
- Dark chestnut-brown coverts with pale fringes
- May have less white on primary bases than adult (varies)

Pomarine Skua (typical)

- Much less variation in colour and tone than in Arctic or Long-tailed; lacks warm-brown tones of many Arctics
- Diagnostic 'double crescent' - pale patch on primary coverts inside main wing-flash
- Head same colour as body, never contrastingly paler
- Undertail-coverts more evenly barred than in Arctic

Arctic Skua (typical)

- Ground colour often warm brown in Arctic (Pomarine invariably darker than this bird)
- Head often contrastingly pale brown (not Pomarine)
- Nape may be reddish in tone (never in others)
- Close view shows head to be streaked (barred in Pomarine)

Long-tailed Skua (typical)

- Ground colour almost always distinctive grey-brown, with 'colder' appearance than tones of Arctic
- Often shows more severe, neat 'black-and-white' barring on underparts than other juvenile skuas
- Undertail-coverts usually clearly barred, even on dark birds (barring may seem absent in darker Arctics)
- Single, small wing-flash

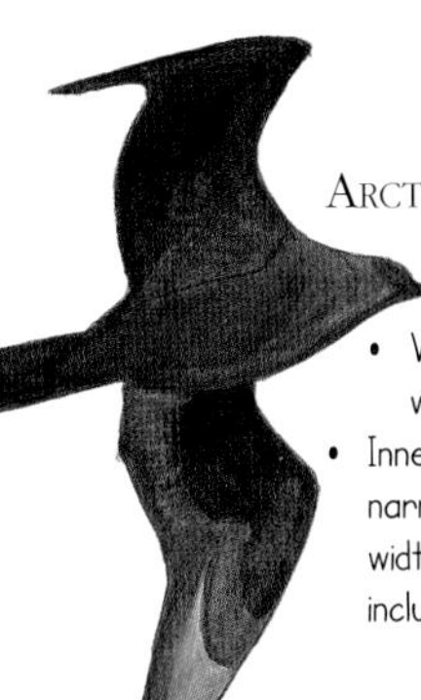

Arctic Skua (dark)

- Outer wing (hand) long
- White flash on wing variable
- Inner wing (arm) fairly narrow - about same width as tail is long (not including projections)

Pomarine Skua (dark)

- Breadth of inner wing greater than distance from trailing edge of wing to beginning of tail streamers (smaller in Arctic and Long-tailed)
- Inner wing (arm) broad - slightly broader than tail is long (not including projections of latter)
- Thicker neck and heavier head than Arctic's
- Broad base to tail

Long-tailed Skua (dark)

- This is an exceptionally dark bird - note juvenile Long-tailed much more variable in colour than adult, which is always pale
- Diagnostically small wing-flash on upper wing - just a couple of primaries white, much less than in other skuas
- Definitively long rear end

Arctic Skua (from above)

- At close range, check for two small, sharp projections at tail-tip

Long-tailed Skua

- Fairly short, thick bill
- Bill approximately half black and half grey (40-50 per cent black)
- Black on bill extends beyond gonydeal angle (c.f. Arctic and Pomarine)

Arctic Skua

- Bill more slender than that of Pomarine and longer than in Long-tailed
- Only tip of bill black, much less than in Long-tailed
- Blue-grey at bill-base less conspicuous than in Pomarine (partly because colour is darker)

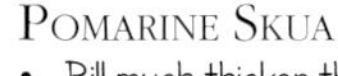

Pomarine Skua

- Bill much thicker than Arctic's, but equally long
- Bill grey with black tip, as Arctic
- Pale colour of bill-base much more noticeable than in Arctic (often at considerable distance), at least partly because it contrasts strongly with dark plumage on head

Juveniles (continued)

Long-tailed Skua (typical)

- Plumage generally 'colder' and greyer in tone than similar Arctic's
- Often pale on nape and sides of head (can be much paler than this)
- Narrow pale band across chest (not in dark individuals)
- Vent pattern always strongly barred against white background
- White fringes to coverts contrast sharply with dark plumage, producing scaly pattern

Arctic Skua (typical)

- Upper breast and head lightly streaked (barred in Long-tailed and Pomarine)
- Primaries usually have clear pale tips (never in Pomarine or Long-tailed), but some very dark Arctics lack feature
- Warm chestnut fringes to coverts

Pomarine Skua (typical)

- In contrast to Arctic, head never streaked, but may be barred or peppered
- Head with distinct grey wash (Arctic always warmer toned, Long-tailed often whiter)
- Strong and close barring on underparts
- Feather fringes often pale (similar to those of Long-tailed) or may have warmer tone

Pomarine Skua (dark)

- Bulkier and more thickset than Arctic and Long-tailed, with heavier head
- Close-set barring on underparts
- Ventral pattern striped black and white/rufous (dark Arctics have no stripes on vent, just dark brown)
- Primaries uniformly brown

BLACK-HEADED GULL

INTRODUCTION Black-headed Gull (*Chroicocephalus ridibundus*) [L 37cm] common resident over much of area. Breeds on freshwater marshes and lakes, and by lagoons and salt marshes; ubiquitous in winter. It is a 'two-year' gull, so plumage sequence is Juvenile – 1st winter – 1st summer – Adult winter – Adult summer (see introduction for terminology).

Ageing Black-headed Gull

Juvenile, June–August

- Anaemic brown markings on head help to make plumage look odd
- Heavy markings on mantle

1st winter

- Brown spots and streaks on wings (especially coverts and tertials)
- Head pattern already similar to adult winter's
- Orange-red legs (often paler than adult winter's)
- Orange-red bill with black tip
- Brown marks on coverts (c.f. adult)
- Black tail-band (c.f. adult)
- White leading edge 'triangle'

1st summer

- Lots of brown on wings, especially tertials (as first winter)
- Brown head mask (as adult), but blotchy and incomplete (c.f. adult)

Adult winter

- Bill often slightly deeper red than first winter's
- Head pattern otherwise as first winter
- Black splodge behind eye and shadow just above eye typical of winter plumage
- No brown blotches on wings or black tail-band

Adult summer

- Deep red bill and legs (deeper than in adult winter or immature stages)
- Smart chocolate-brown hood; white eye-ring
- White tail
- Clearly defined white triangle on leading edge of wing typical of adult

COMMON GULL

INTRODUCTION Common Gull (*Larus canus*) [L 43cm] widespread breeding bird right into the Arctic, on marshes, by lakes and on moorland, often far inland; widespread in winter in many habitats, from coasts to farmland and fields.

Common Gull is a 'three-year' gull, so plumage sequence to adulthood is Juvenile – 1st winter – 1st summer – 2nd winter – 2nd summer – Adult winter – Adult summer (see Introduction for terminology).

Ageing Common Gull

JUVENILE

- Head mottled brownish
- Pale fringes to coverts give scaly pattern

1ST WINTER

- Head evenly covered with speckles
- Bill pink with black tip
- Legs flesh coloured
- Neat cold-grey mantle contrasts with brown wing-coverts. Sheer tidiness, with well-ordered grey-and-brown panels, is good marker - most first-winter gull species look messier

2ND WINTER

- Smaller white spots (mirrors) on folded wing-tip than in adult winter
- Less obvious tertial crescent than in adult
- Thicker dark band on bill than in adult

ADULT WINTER

- Pleasing, bluish-grey 'cold' colour to mantle
- Yellow legs
- Yellow bill with thin dark band
- Speckles on head (lost from February, leaving clean white head)

JUVENILE
- Blotchier than first winter
- Greater wing-coverts greyish-brown; not prominent

1ST WINTER
- Wing basically looks brown, but contains grey panel made up by greater wing-coverts
- Head and body finely mottled brown, with no hint of face-mask
- Black tail-band

EARLY 2ND WINTER
- Small smudges of black at angle of wing (primary coverts) - absent in adult
- Largely black wing-tips, with small mirrors (c.f. adult winter)
- Legs greenish or blue-grey

LATE 2ND WINTER
- Similar to early second winter, but black on wing-coverts begins to wear
- Greenish bill with black band near tip (as in adult)

ADULT WINTER
- Flies with full, smooth wingbeats, giving fluent flight action
- Wing-tips have big white blobs, contrasting with black just inside - excellent indicators of adult
- White tips to inner primaries (c.f. second winter)

ADULT SUMMER
- Rounded pure white head (c.f. adult winter)
- Dark eye gives 'peaceful' expression
- Yellow bill sometimes retains grey subterminal band
- Greenish-yellow legs
- Black wing-tips with large white mirror more distinctive than in other gulls

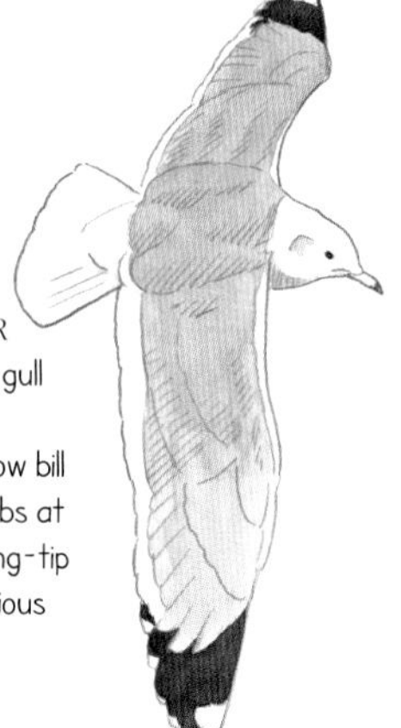

ADULT SUMMER
- Neat, compact gull with full wings
- Relatively narrow bill
- Large white blobs at end of black wing-tip - especially obvious on underside

SMALL GULLS

INTRODUCTION This section compares the smaller gulls plumage by plumage. For general information on these species see relevant sections.

ADULT SUMMER GULLS WITH HOODS

MEDITERRANEAN GULL

- Smart black hood (chocolate-brown in Black-headed)
- White crescents around eye
- White wing-tips
- Bright red bill with yellow tip and black subterminal band
- Paler above than Black-headed and, especially, similar-sized Common

LITTLE GULL

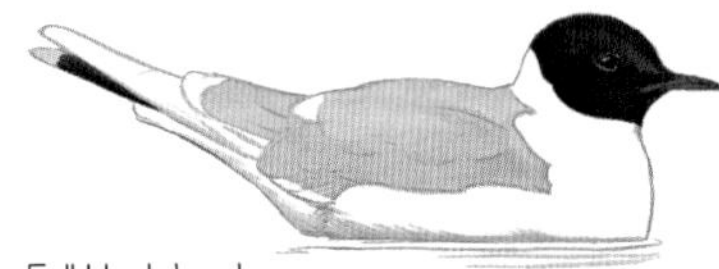

- Full black hood
- Very small (half size of Mediterranean) and compact
- Red-brown bill, but looks black
- No white eye-crescents

- Rather rounded white wing-tips
- Flies erratically, with fast wingbeats and jinking flight
- Blackish underwings unique

BLACK-HEADED GULL

- Rather narrow and pointed, deep red bill
- Smart chocolate-brown hood and white nape
- Black wing-tip with very small white mirrors
- Rather pale grey back and mantle

- Sharper wing-tips than Mediterranean's
- Outer wing pattern diagnostic: clean white triangle bordered by narrow black wing-tip and grey inner primaries - great field mark at all ages
- Darker grey on back than Mediterranean

Adult winter

Black-headed Gull

- Distinctly sharp-pointed wings - would 'prick a balloon', unlike wing-tips of Common or Mediterranean
- Wingbeats quick and, at times, seem to flicker; very agile species that fights with ducks for bread
- White leading edge to outer wing from below

- Long, thin, fine-tipped bill
- Smudged-ink marks behind and above eye different in shape from Mediterranean's; black spot leaks up towards cap
- Pale-grey mantle, much paler than in Common, but darker than in Mediterranean
- Black wing-tips with very small white mirrors

Mediterranean Gull

- Flies with rather stiff, fast, shallow wingbeats - very distinctive and can be distinguished at long range
- Wings look all white, almost like a Little Egret's
- Shorter, broader, blunter wings than in Common or Black-headed

- 'Bruise' behind eye
- Large, thick, blob-tipped deep red bill stands out at least as much as white wing-tips
- Flatter back and squarer head than in Black-headed
- Clean white wing-tips

Little Gull

- Very small and dainty. Flies with fast, full wingbeats, and hunts by flying low over water, often dipping down suddenly to pick food from surface (similar to a marsh tern)
- Wings mainly pale grey, but with neat white trailing edge that also forms wing-tip (distinctive)
- Two-toned underwings, grey in front and blackish towards back
- Black spot behind eye (as Black-headed)
- Black crown (unique)

Herring Gull (*argenteus*)

- Low crown and 'frowning' expression - similarly plumaged (but much smaller) Common has gentle, doe-eyed expression
- Heavy yellow bill with orange spot
- Pink legs
- Very pale, silvery back (see Common)

Common Gull

- Dark eye
- Grey speckling on head - feature lacking in smaller gull species
- Greenish bill with black subterminal band (often 'Z'-shaped)

Common Gull

- Fluent flight action with full, smooth wingbeats
- Big white blobs at wings tips, contrasting with black just inside
- White tips to inner primaries, giving very different wing pattern from that of Black-headed
- Wings darker grey above than in Black-headed

1st winter in flight

Common Gull

- Head and body finely mottled brown, with no hint of face-mask
- Brown wings not very contrasting
- Underwing speckled brown; not as clean as that of Mediterranean
- Smaller, thinner bill than in Mediterranean

Kittiwake (juvenile/1st winter)

- Very distinctive flight style, with quick, shallow wingbeats on stiff wings
- Similar black 'W' pattern on upperwings to Little's
- Outer edge of upper wing pure white (c.f. Little Gull)
- Large black saddle on shoulder (c.f. others)

Little Gull

- Much smaller than Black-headed, with dainty, lively flight
- Wings short but quite pointed (wing-tips of adults more rounded)
- Dark 'W' across wings distinguishes it from Black-headed
- Dark wing-tips - amount varies individually

Mediterranean Gull

- Dark, 'bruised' ear-coverts making mask
- Pale grey panel in middle of wing contrasts strongly with leading and trailing edges (c.f. Common)
- Whitish underwing
- Small white spots on wing-tip

Black-headed Gull

- Diagnostic white 'wedge' on leading edge of outer wing (in all ages)
- Narrow black edge to outer wing (no mirrors)
- Well-defined black tail-band
- Dark patch at tip of underwing

1st winter

Black-headed Gull

- Bill dark orange-red with black tip
- Dark orange-red legs (paler than Mediterranean's)
- Very pale grey mantle
- Head white, not speckled and marked only with well-defined black spots on ear-coverts and around eye; colour of both may 'leak' upwards towards crown
- Dark spots on wing-coverts

Common Gull

- Distinctly chunkier than Black-headed, with rounder head and thicker bill than in Black-headed
- Neatly proportioned, with long, pointed wings
- Head evenly covered with speckles
- Bill pink with black tip
- Neat, even lines of brown-centred coverts

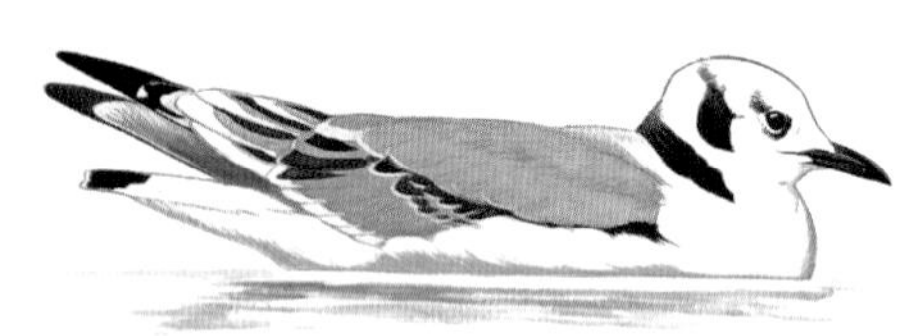

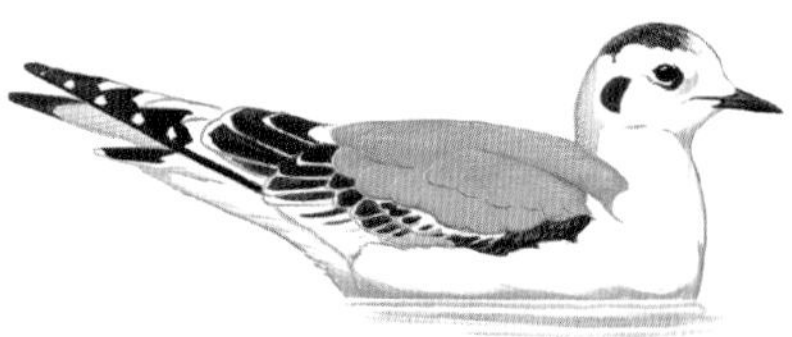

Kittiwake (juvenile/1st winter)

- White cap (c.f. Little)
- Black cheek spot (smaller than in Black-headed)
- Grey mid-wing-panel
- Bold black colar on hindneck

Little Gull

- Black spot behind eye (as Black-headed)
- Black crown
- Small and compact
- Short, fine black bill

2nd winter

Mediterranean Gull

- Closely resembles adult winter except for extent of variable amount of black on outer five primaries (may be much less than this)
- Bill often dull orange with black band near tip (may be redder)

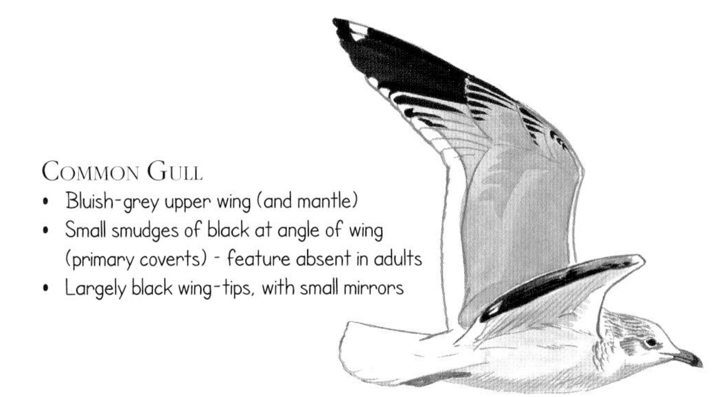

Common Gull
- Bluish-grey upper wing (and mantle)
- Small smudges of black at angle of wing (primary coverts) - feature absent in adults
- Largely black wing-tips, with small mirrors

UVENILE

Mediterranean Gull (August)

Contrasting whitish head (no sign of distinctive dark ear-coverts seen in first- and second-winter birds)

Strong, neat, scalloped back

Contrasting white underparts

Long black legs

Distinctive pale-grey bar made by greater wing-coverts

Common Gull
- Weaker, thinner bill than Mediterranean's
- Browner and less contrasting than Mediterranean juvenile
- More rounded head and 'gentler' expression than Mediterranean's
- Head mottled brownish

Black-headed Gull
- Extensive brown on head
- Heavy markings on mantle

Little Gull
- Black outer primaries and black mid-wing-panel make 'W' zigzag pattern across two wings (retained through first winter)
- Bold black cap
- Large dark patch on 'shoulder'
- Very distinctive scaly appearance to blackish mantle, with white feather fringes

MEDITERRANEAN GULL

Mediterranean Gull (*Larus melanocephalus*) [L 38cm] scarce and localized breeding bird, in salt marshes and lagoons; more numerous in winter in a variety of habitats, usually coastal

AGEING MEDITERRANEAN GULL

JUVENILE

- Dark above with bright white feather fringes giving scalloped effect
- Clean white underneath
- Clear band of grey on greater wing-coverts
- Oddly plain face

JUVENILE (FLYING)

- Clean grey greater wing-covert band
- Dark mantle (c.f. first winter)

JUVENILE MOULT–1ST WINTER

- Scalloped pattern fades leaving plain grey back

1ST WINTER (FLYING)

- Narrow tail-band
- White patches on primaries
- Neat grey mid-wing band

1ST WINTER

- Legs blackish to reddish
- Bill black, at end of winter showing reddish base
- Distinctive dark wedge-shaped 'mask' on head, behind eye and up to crown
- Small dark brown mid-wing panel

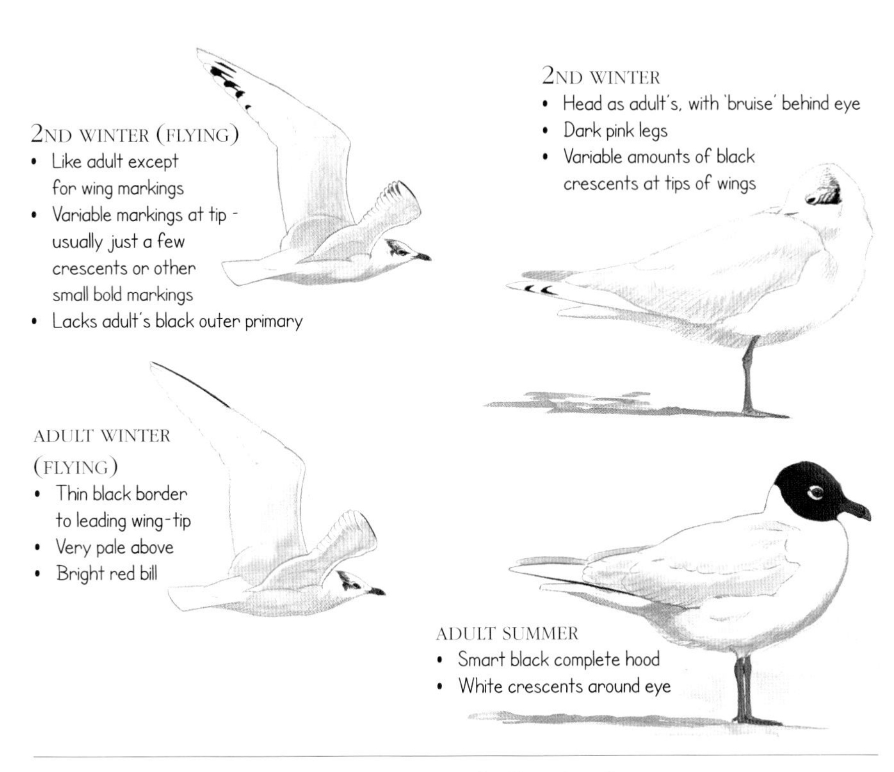

HERRING, YELLOW-LEGGED & CASPIAN GULLS

INTRODUCTION Herring Gull (*Larus argentatus*) [L 60cm] breeds abundantly over much of the region, on cliffs and dunes, and by lakes; mainly coastal. Winters widely, mainly by the sea. Yellow-legged Gull (*Larus michahellis*) [L 57cm] breeds along the French Atlantic coast and widely in the Mediterranean; wanders north (as far as southern Britain to Denmark) from late summer onwards. Caspian Gull (*Larus cachinnans*) [L 62cm] breeds Eastern Europe by inland lakes and rivers; uncommon non-breeding visitor to the Baltic westwards to Britain.

HERRING GULL RACES

argenteus breeds in Britain, Iceland, France and the North Sea coast east to about Denmark (intergrades occur from northern France eastwards).

argentatus breeds in northern Europe and is a winter visitor further south.

Adult winter standing

Herring Gull (*argenteus*)

Features of Herring Gull

- Yellow bill with orange spot - orange confined to lower mandible, unlike in Yellow-legged
- Pale, beady eyes (cold, 'fish-eyed' expression) - looks fierce
- Grey, slightly bluish-tinged upperparts
- Pink legs
- Somewhat truncated rear end (c.f. Yellow-legged)
- In winter, usually much brown specking about head and neck, making it look rather messy (c.f. Yellow-legged and Caspian)

Features of *argenteus*

- Very pale, silvery back
- Much black on wing-tip; outer primary P10 has white mirror

Herring Gull (*argentatus*)

Differences from *argenteus*

- Larger and heavier
- Slightly darker upperparts, closer to those of Yellow-legged
- Wing-tip (made by P10) usually white
- May show more speckling on winter head than *argenteus*

Yellow-legged Gull

Differences from Herring

- Slightly larger and more powerful, with rugged head and still heavier bill, longer wings and leaner body (= 'Herring in the front and Lesser Black-backed behind')
- Darker grey mantle and wings, lacking any blue cast
- Long yellow legs
- Large orange spot on bill leaks on to upper mandible
- Orange to blood-red orbital ring (makes eyes look dark)
- More black on wing-tip, with smaller mirrors

Caspian Gull

Differences from Yellow-legged

- Yellowish legs with pinkish cast - not as bright
- Small, beady dark eye
- Long and relatively thin, parallel-edged bill - may have greenish cast
- Much more slender body (also than Herring)
- Paler grey back (closer to Herring)

Adult summer in flight

Herring Gull (*argenteus*)

Differences from *argentatus*

- Pale grey, lighter than *argentatus*
- More black on wing-tips
- Very tip of P10 has less white (if any)

Differences from Yellow-legged

- Much less black and more white on wing-tips than in Yellow-legged
- Paler above than Yellow-legged

Herring Gull (*argentatus*)

- White triangle on wing-tips caused by white P10 and large inner mirror on P9
- More obvious and better defined white tips to inner primaries

Yellow-legged Gull

- Small, narrow mirrors
- Very little white and a lot of black on wing-tip (see Herring)
- Distinct white trailing edge to inner wing
- Dark grey upperparts

Silhouettes

Yellow-legged Gull

- Plumper than Caspian
- Bill thick with obvious gonydeal angle
- More evenly proportioned than Caspian
- More obvious tertial step than in Caspian

Caspian Gull

- Bill long and parallel sided
- Peculiar small, pear-shaped head with sloping forehead
- Longer winged than Yellow-legged, and rear end looks generally more attenuated
- Long legs
- Much longer neck than in Yellow-legged or Herring

Juvenile and 1st winter

Yellow-legged Gull (juvenile)

- Noticeably contrasting, with white head and underparts and dark upperparts
- Thick black bill
- Dark mask
- Pink legs (no help for separating from Herring)
- Outer greater wing-coverts paler than inner ones (equivalent Herring does not show this)

Herring Gull (juvenile–1st winter)

- Head usually more uniformly dark than that of Yellow-legged or Caspian, and with fairly dark mask around eye
- Well-marked tertials with dark branched centres and pale edges - 'oak-leaf' pattern (relatively plain in Yellow-legged and Caspian)
- Neatly patterned greater wing-coverts (see others)

Yellow-legged Gull (1st winter)

- Distinctively white head a good indicator (as opposed to Herring)
- Often distinctive varied tones to body: brown, white, greyish and black
- Dark tertials with narrow white fringes (Herring Gulls of similar age have barred 'oak-leaf' pattern, very different, and tertials are browner)
- Dark mask and streaks on nape
- Black bill (with slightly pink base, showing less than equivalent Herring)

Caspian Gull (1st winter)

- Scaly scapulars with some anchor-shaped dark markings (more prominent in late winter)
- Tertials dark with fairly broad whitish tips (cleaner but narrower tips in Yellow-legged, oak-leaf patterned in Herring)
- Mainly white forehead with smudging around eye (no mask, unlike in Yellow-legged or Herring)
- Whitish underparts
- Brown hindneck contrasts with white head

Herring Gull (1st winter)

- Rump heavily peppered with dark spots
- Inner primaries paler than outer ones, producing prominent pale 'window' on outer wing
- Single dark bar near trailing edge formed by secondaries

Yellow-legged Gull (1st winter)

- Clean-cut, well-defined black tail-band
- Darker greater wing-covertsthan those of Herring
- Whitish rump with only a few dark spots (c.f. Herring)
- Whiter head than Herring's

2nd winter

Yellow-legged Gull

- Whitish head with fine streaks on crown and nape
- Matures more quickly than Herring, with more grey on mantle than equivalent age
- Distinctive rectangle made by dark-barred greater wing-coverts (does not show in Herring)
- Darker eye than equivalent Herring's
- Dark mask around eye stands out in white head

Herring Gull (*argenteus*)

- Shows limited grey on mantle compared with Caspian and Yellow-legged
- Brownish-patterned greater wing-coverts do not make obvious rectangle
- Median wing-coverts not grey
- Pale eye begins to appear

Yellow-legged Gull

- Retains well-defined tail-bar and white rump Equivalent stage of Herring shows little or no grey on upperwing

Caspian Gull

- Blotchy and indistinct dark markings on lesser wing-coverts
- Dark shawl down nape
- Upperparts already largely grey
- Darker tertials than in Yellow-legged

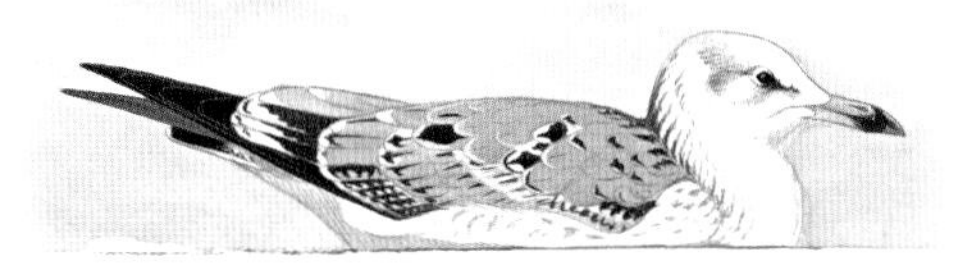

GREAT & LESSER BLACK-BACKED GULLS

INTRODUCTION Great Black-backed Gull (*Larus marinus*) [L 70cm] widespread breeding bird on coasts, especially cliffs and islands; winters in many maritime habitats, less common inland. Lesser Black-backed Gull (*Larus fuscus*) [L 54cm] widespread breeder on dunes, moorland and islands, but not usually on cliffs; summer visitor to the far north. Winters in a variety of habitats, more often inland than Great Black-backed.

Great Black-backed Gull is the world's largest gull. All individuals are larger than Herring Gull, while most Lesser Black-backed Gulls are smaller than Herring Gulls.

Lesser Black-backed Gull Races

graellsii Breeds in the Britain, Ireland, France and Spain, and winters from southern England southwards along the seaboard to the Mediterranean.

intermedius Breeds in Norway, Denmark and Kattegat, and winters widely in Western Europe. Darker grey back than *graellsi*, back still contrasts with black wing-tips.

fuscus ('Baltic Gull') Breeds around the Baltic, winters mainly in East Africa.

Breeding adult

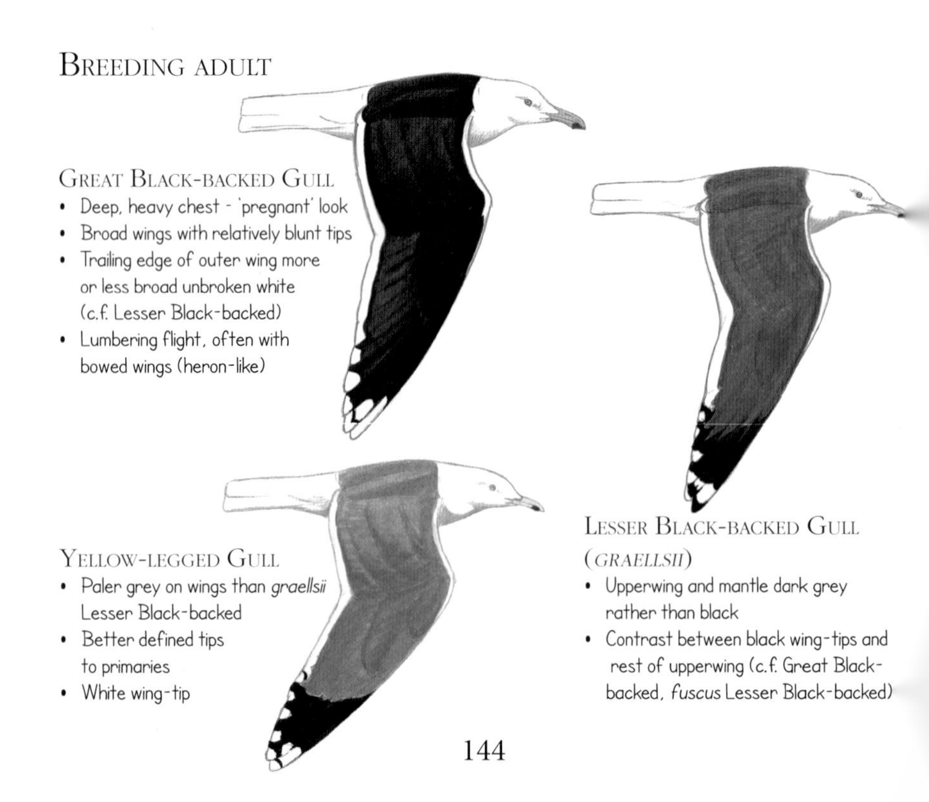

Lesser Black-backed Gull (*fuscus*)

- Yellow legs (paler in winter)
- Far smaller than Great Black-backed; many slightly smaller than Herring
- More rounded head and gentler expression than brutish Great Black-backed's
- Much narrower bill than that of Great Black-backed, without such obvious gonydeal angle; same colour scheme as that of Great Black-backed and Herring
- Looks neat and well proportioned in contrast to 'lumpy' Great Black-backed
- Elegant, tapered rear end with long, pointed wings

Great Black-backed Gull

- Flat forehead and intimidating expression, exacerbated by small eye set far back on head
- Enormous thick bill with blade-like edge to lower mandible, the 'gonydeal angle'; bill tip looks heavy
- Long, pale flesh-coloured legs
- Mantle always deep black
- Comparatively short wings, giving stumpy rear end

Lesser Black-backed Gull (wing, *fuscus*)

- Smooth, easy flight action compared with that of Great Black-backed; more buoyant and gull-like
- Trailing edge of outer wing not continuous white; at best a few white spots

Overall, much less white on wing-tips than in Great Black-backed

Narrower wings than in Great Black-backed, with longer hand and more pointed tip

Lesser Black-backed Gull (adult winter, *graellsii*)

- Paler back than in *intermedius/fuscus*
- Grey contrasts with black wing-tips
- Intensely yellow legs
- Heavily streaked head in winter (as with Herring)

Yellow-legged Gull

- Heavier build than that of Lesser Back-backed, with thick bill; similarly long wings
- Slightly bigger mirrors on wing-tips
- Much paler mantle than in any black-backed

1st winter

Great Black-backed Gull

- Entirely black bill, broad, long and with obvious gonydeal angle
- Very white head
- More contrasting and well-patterned scapulars and coverts than in Herring or Lesser Black-backed; never looks uniform, more black and white
- Broad white edges to tertials

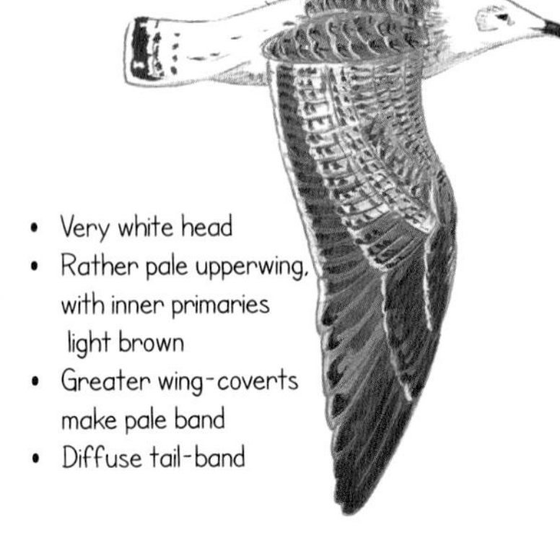

- Very white head
- Rather pale upperwing, with inner primaries light brown
- Greater wing-coverts make pale band
- Diffuse tail-band

Herring Gull

(juvenile–1st winter)

- Slightly thicker bill than that of Lesser Black-backed - both species' bills may show pink bases
- Larger than Lesser Black-backed, with flatter crown and more angled hind-neck
- Tertials barred, with dark fish-bone pattern to centres and pale buff edges

- Outer primaries dark, inner primaries paler brown (as Great Black-backed, but not Lesser Black-backed)
- Rump has more barring than that of Lesser Black-backed, rendering contrast from narrower tail-band less obvious

Lesser Black-backed Gull

(juvenile–1st winter)

- Dark tertials with narrow white edges
- Darker backed than Herring, with obvious contrast between upperparts and underparts
- Sharper wing-tips than Herring's

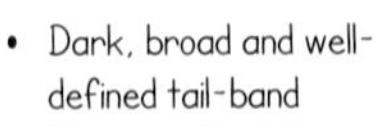

- Dark, broad and well-defined tail-band
- Rump mainly white, providing sharp contrast to dark tail-band
- Outer primaries entirely dark, giving whole wing monotone appearance
- Darker backed than Herring

GLAUCOUS & ICELAND GULLS

Glaucous Gull (*Larus hyperboreus*) [L 65cm] breeds in Iceland (Iceland Gull does not) and is an uncommon winter visitor (October–April) to the Atlantic and North Sea coasts down to northern France. Iceland Gull (*Larus glaucoides*) [L 56cm] breeds in Greenland, with small numbers moving south in winter to the area.

Iceland and Glaucous Gulls do not have a first-winter plumage because they do not moult their juvenile plumage in autumn. Hence page 148 refers to juvenile/first winter.

Silhouettes

Glaucous Gull

- Big and bulky - some almost as large as Great Black-backed
- Heavy chested
- Longer neck than Iceland's
- Short primary projection (distance from tertials to wing-tip)
- Wings do not project far beyond tail (distance no greater than length of bill)
- Pronounced tertial step

Iceland Gull

- Smaller than Glaucous; slightly smaller than Herring
- Shorter legs than Glaucous's
- Long primary projection (long, elegant wings project well beyond tail - a distance much greater than length of bill)
- Full, deep pigeon-breast

2nd winter

Glaucous Gull

- Paler than first winter
- Rather blotchy - like a 'painted-over' first winter
- Trace of pale grey appears on mantle, especially in late second winter
- Both species often show dark mask by eye at this stage

Iceland Gull

- Exhibits obvious contrast between base and tip of bill, unlike in first winter, increasing similarity to Glaucous
- Both species acquire pale eyes at this age
- May show green tinge to bill (more so in third year, and sometimes in adult - never in Glaucous)

Juvenile/1st winter

Iceland Gull

- Wings much more pointed, with relatively longer hand
- Hand and arm of equal length, so looks neat and balanced
- Agile flight for a large gull; capable of snatching food from water's surface (Glaucous cannot do this)

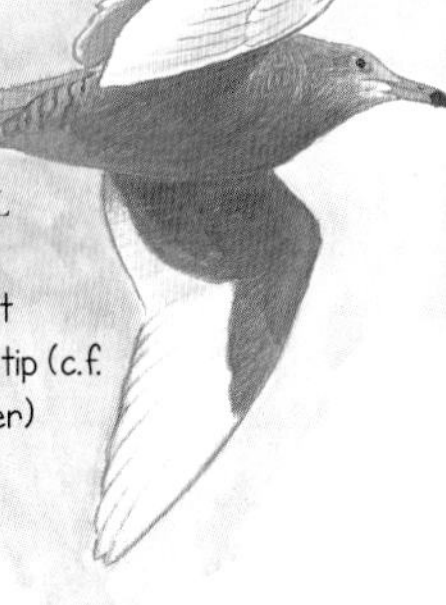

Glaucous Gull

- Outer wings often appear translucent
- Pink bill with black tip (c.f. Herring first winter)
- Broad wings

Iceland Gull

- Rounded head and dark eye give 'friendly' expression (= 'Common Gull in large gull's body')
- Slim and flat backed
- Overall 'cappuccino'-coloured plumage - white with complex brownish speckling

Glaucous Gull

- Heavy and bulky with rounded back
- Bulge at tertials
- Notably stout ended, in contrast to Iceland's slim and pointed rear end
- Rather small eye for head

Iceland Gull

- Gentle expression
- Rounded crown and relatively steep forehead
- Much narrower and shorter bill than Glaucous's
- Contrast between pink base and black tip not as clear as in Glaucous - black shades into pink

Herring Gull (juvenile)

- Dark face-mask
- All-dark bill (by late first winter, some have small trace of pink at base)
- Longer bill than Iceland's, and with more obvious gonydeal angle
- More wedge-shaped head than Iceland's

Glaucous Gull

- Long, parallel-edged bill with obvious cutting gonydeal angle
- Clearly contrasting pink and black on bill
- Rather wedge-shaped head with flat crown and angled rear crown
- Long, gently sloping forehead

LITTLE GULL

Little Gull (*Hydrocoloeus minutus*) [L 26cm] localized breeding bird of the North Sea and Baltic coasts, although it breeds in freshwater marshes. Coastal in winter; encountered widely on passage, on sea and inland waters.

May recall marsh terns in feeding technique: it flies at a consistent height above water except when it suddenly dips down to snatch an item from the surface. It also patters water like a storm-petrel and may hover briefly.

Ageing Little Gull

juvenile (June–September)

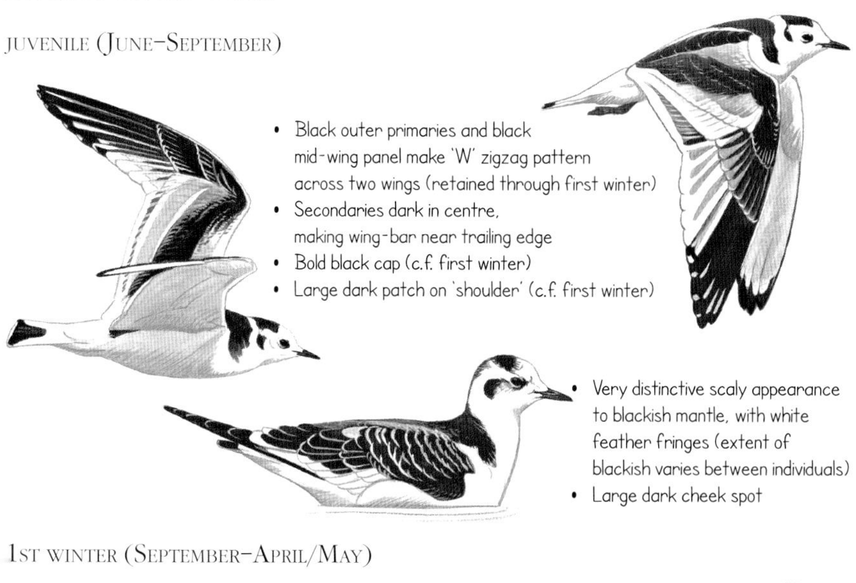

1st winter (September–April/May)

Ageing Little Gull (continued)

1st winter (continued)

- Pink legs
- Black spot behind eye (as adult winter)
- Black 'shoulder-bar'
- Dark centre to crown

1st summer (April/May–August)

- Some individuals attain partial black hood (often less extensive than this)
- Orange legs
- Immature wing pattern retained, but looks worn
- Often rosy tinge to breast and/or belly
- Some individuals acquire white tail

- Partial dark hood
- Worn tail-band (but some still remains)
- Very worn juvenile wing pattern
- Underwing still grey (c.f. adult)

2nd winter (September–March/April)

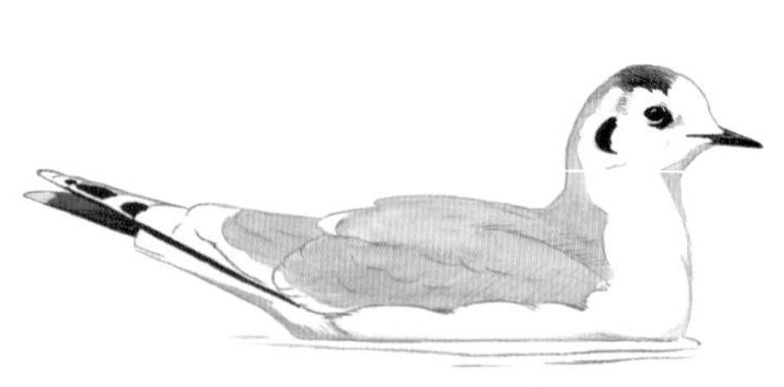

- Face as adult's, with dark crown and small dark cheek spot
- Small, thin bill and rounded head give gentle, almost 'cute' appearance
- Wing-tip pattern at rest shows just a little black; adult pure white

- Wing mainly grey with variable black markings at tip
- On underwing, pale grey coverts contrast with blackis band along primaries and secondaries (c.f. first winter

ADULT SUMMER

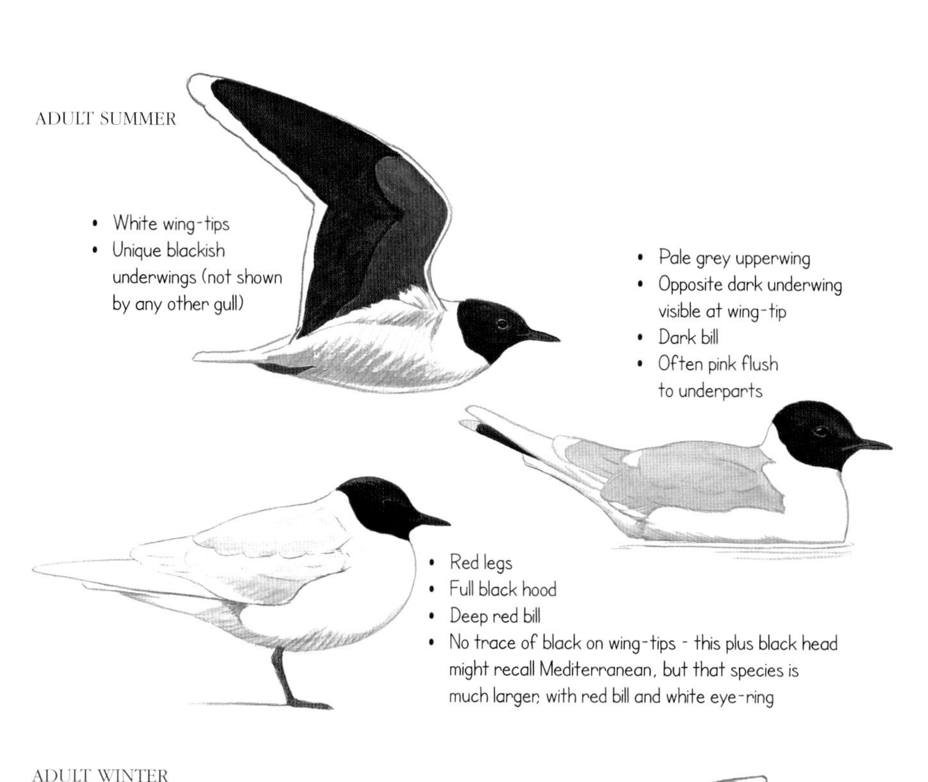

ADULT WINTER

- Wings short, broad and rounded, with rounded effect perhaps accentuated by white wing-tip
- Wings mainly pale grey, but with neat white trailing edge that also forms wing-tip (distinctive)
- Two-toned underwings, grey in front and blackish towards back
- Grey upperwing/dark underwing contrast very distinctive in flight
- Dark pink legs

KITTIWAKE, LITTLE & SABINE'S GULLS

INTRODUCTION Black-legged Kittiwake (*Rissa tridactyla*) [L 40cm] is a common and numerous ocean-going gull that breeds on cliffs throughout much of the region. Sabine's Gull (*Xema sabini*) [L 33cm] is an Arctic breeding species (nearest colonies in Svalbard), which migrates down the mid-Atlantic and can be brought towards land by westerly gales. It is rare but regular off western Britain and Ireland, and the Bay of Biscay, in autumn.

These gulls share a similar wing pattern in the immature stages, with dark inner and outer bars making a 'W' shape in flight. They also share a pelagic habitat – that is, they are generally found far out to sea in the non-breeding season.

Juvenile/1st winter

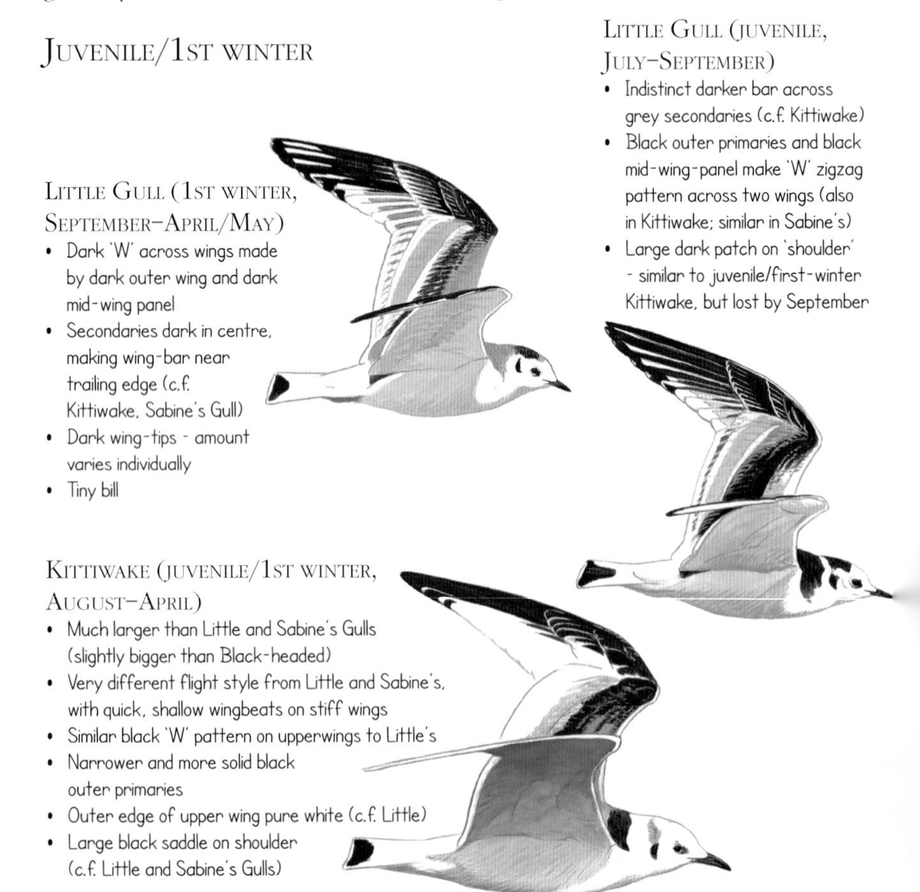

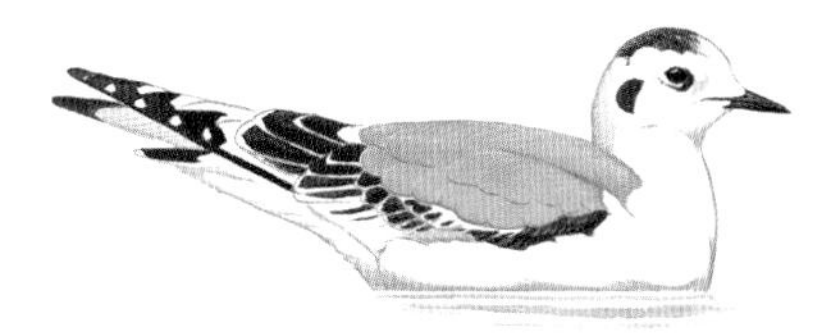

LITTLE GULL (1ST WINTER)

- Rounded head and 'cute' expression
- Very small, narrow bill
- Black cap (c.f. Kittiwake, Sabine's Gull)
- Black spot behind eye (as Kittiwake, c.f. Sabine's Gull)

KITTIWAKE (JUVENILE/1ST WINTER)

- Much larger than Little (almost twice size)
- Longer body and not as compact or buoyant as Little
- Large black shoulder-patch (confusion possible with juvenile and moulting Littles)
- Grey mid-wing-panel

LITTLE GULL (ADULT WINTER)

- Narrow black bill
- Dark grey crown
- Large black spot behind eye
- Rounded head

SABINE'S GULL (JUVENILE)

- Fairly long black bill
- Greyish-brown crown, ear-coverts, nape and neck sides
- Scaly mantle
- No black on head plumage

KITTIWAKE (1ST WINTER)

- Small black spot behind eye
- Large black patch on nape diagnostic
- White crown
- Grey mantle

SABINE'S GULL (JUVENILE, JUNE–DECEMBER)

- Diagnostic 'three triangles' on wing, of white, black and brown-grey
- Broad white trailing edge to wings extends in triangle to bend of wing
- Scaly greyish-brown coverts and mantle
- Shallow fork to tail
- Small head, and wings look 'too large for body'.

KITTIWAKE (JUVENILE/1ST WINTER, AUGUST–APRIL)

- Black panel across mid-wing (as Little, c.f. Sabine's)
- White secondaries and inner primaries (as Sabine's)
- Black nape-patch

SEA TERNS

INTRODUCTION Common Tern (*Sterna hirundo*) [L 36cm] common summer visitor April–September throughout most of the region, breeding on islands, beaches, rivers and freshwater wetlands. Arctic Tern (*Sterna paradisaea*) [L 36cm] arrives a little later, late April, and is more northerly, breeding from Britain north to the Arctic, in similar habitats to Common but also on tundra. Roseate Tern (*Sterna dougallii*) [L 34cm] rare summer visitor May–September to a few low-lying coastal rocky islets off France, Britain and Ireland; rare migrant elsewhere. Sandwich Tern (*Thalasseus sandvicensis*) [L 40cm] common summer visitor March–October to coastal islands, sand-spits and beaches in Britain, France, the North Sea coast and some Baltic islands. Gull-billed Tern (*Gelochelidon nilotica*) [L 38cm] mainly summer visitor to southern Europe, but also breeds on lakes and rivers in northern Germany and Denmark. Little Tern (*Sternula albifrons*) [L 23cm] summer visitor to sheltered coasts and rivers; widespread but local.

Terns differ from gulls in having much slimmer and more pointed wings, much longer, frequently forked tails, typically slim, pointed bills and relatively short, weak legs. They rarely swim, unlike gulls.

Adult summer in flight

Common Tern

- Slightly paler above than Arctic
- Dark 'wedge' formed by darker outer primaries compared with paler inner primaries
- Wings look centrally placed on body
- 'Longer,' more protruding head and neck than Arctic's

Arctic Tern

- Slightly smaller body than Common
- Shorter bill than Common's
- Shorter wings but much longer tail than Common's; wings may look a little forwards-set, making bird look rather swallow-like

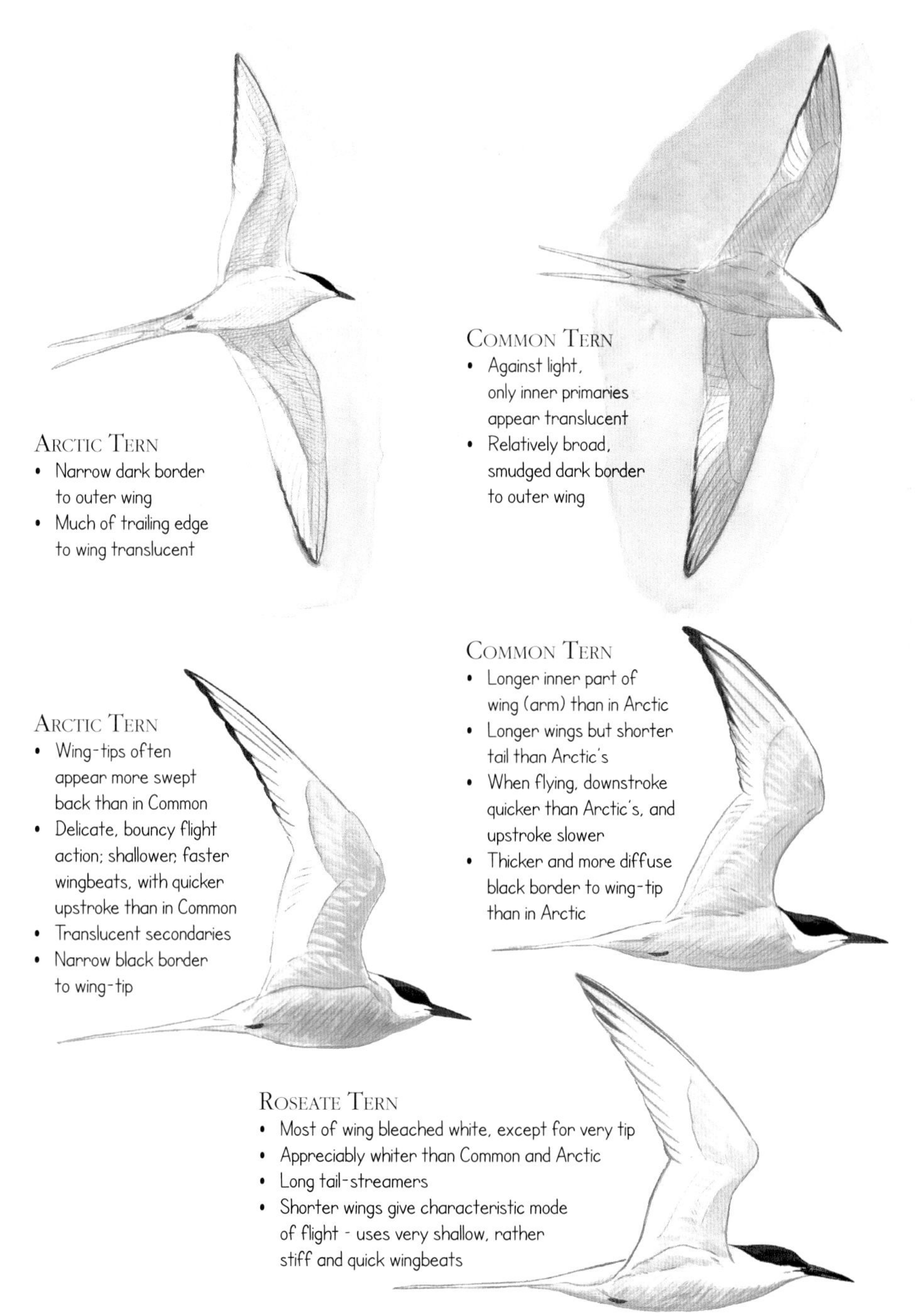
Arctic Tern
• Narrow dark border to outer wing
• Much of trailing edge to wing translucent
Common Tern
• Against light, only inner primaries appear translucent
• Relatively broad, smudged dark border to outer wing
Arctic Tern
• Wing-tips often appear more swept back than in Common
• Delicate, bouncy flight action; shallower, faster wingbeats, with quicker upstroke than in Common
• Translucent secondaries
• Narrow black border to wing-tip
Common Tern
• Longer inner part of wing (arm) than in Arctic
• Longer wings but shorter tail than Arctic's
• When flying, downstroke quicker than Arctic's, and upstroke slower
• Thicker and more diffuse black border to wing-tip than in Arctic
Roseate Tern
• Most of wing bleached white, except for very tip
• Appreciably whiter than Common and Arctic
• Long tail-streamers
• Shorter wings give characteristic mode of flight - uses very shallow, rather stiff and quick wingbeats

Adult summer

Common Tern

- Longer and more downcurved bill than Arctic's; orange-red with black tip
- No contrast between underparts and cheek
- Tail streamers do not protrude beyond wing-tips at rest

Arctic Tern

- Bill shorter than Common's; blood red without black tip
- Tail sticks out well beyond wings at rest
- Overall slightly darker plumage than Common's, so white cheek contrasts with grey neck
- Shorter legs than Common's

Roseate Tern

- Very long bill; black with some red at base (more later in summer)
- Slight or pronounced rosy flush to breast
- Very pale overall
- Longer tail-streamers than in other terns, sticking out well beyond wing-tips

Juvenile/1st winter

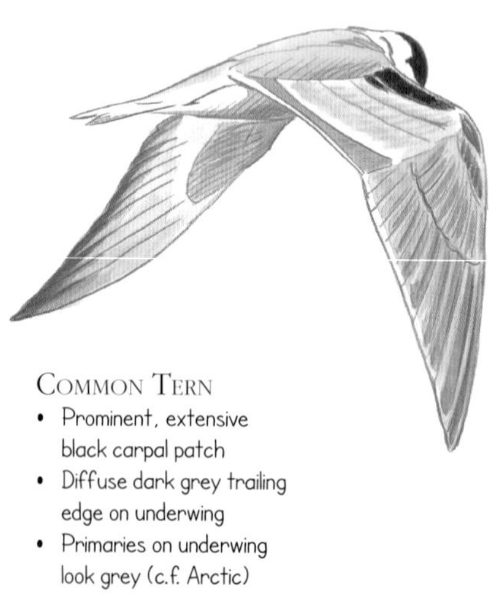

Common Tern

- Prominent, extensive black carpal patch
- Diffuse dark grey trailing edge on underwing
- Primaries on underwing look grey (c.f. Arctic)

Arctic Tern

- Wing becomes lighter in colour from front to back
- White trailing edge to secondaries - diagnostic feature recalling similar plumage on immature Kittiwake and even Sabine's Gull
- Underside very white
- White rump

COMMON TERN
• Secondaries grey, tipped white, forming dark trailing edge to wing
• Distinct gingery tone to upperparts (feature retained well into summer)
• Grey outer-tail feathers
• Wings tips diffusely dull grey
ARCTIC TERN
• Much whiter and neater than Common - noticeable even at great distance
• White outer-tail feathers
• Carpal bar much less prominent, and less well defined than Common's
• Less obvious buff edges to coverts and mantle, and colour fades more quickly than in Common
SANDWICH TERN (1ST WINTER)
• Strong pattern on upperwing, with dark primaries and primary coverts; other coverts have remnants of juvenile pattern
• 'Receding hair' hindcrown
• Thin, dark bill with slight hint of yellow tip
• Black tips to tail
COMMON TERN
• Plenty of orange-red coloration at bill tip (all summer and into autumn)
• Fairly flat crowned
• Extensive white on forehead
ARCTIC TERN
• Mainly dark bill - red base at first, but all black by August-early September
• Darker and much shorter legs than Common's - can look almost legless
• Higher crown than in Common, with small, rounded head
• More black on head, especially behind eye

Roseate Tern (juvenile)

- Dark forehead gives almost complete black cap
- Scaly pattern on upperparts easily separates bird from Common and Arctic Terns (similar to Sandwich Tern juvenile, but much smaller)
- Darker legs than juvenile Common or Arctic's

Gull-billed Tern (juvenile)

- Unusually long legs for tern
- Beginnings of dark mask around eye (seen in non-breeding season, c.f. other terns here)
- Small dark crescents on tertials and wing-coverts - Sandwich juvenile has strongly scaly pattern like Roseate
- Ochre wash to crown, nape and coverts

Little Tern (1st winter)

- White secondaries
- Minute for tern, like gigantic butterfly at times; size alone usually identifies species
- Strong dark carpal bar and grey central panel
- Black bill with hint of yellowish tip
- Compare also with marsh terns (pages 159-160)

Adult summer with black bill

Gull-billed Tern

- Unusual, thick and heavy bill for tern, reflecting diet of insects, crustaceans and even vertebrates such as lizards
- Hawks insects over dry land (c.f. Sandwich); does not splash-dive
- Broad-based wings and short tail
- Dark trailing edges to outer wings
- Greyer rump than Sandwich's

Sandwich Tern

- Flies with strong, full and heavy wingbeats
- Long, thin bill with yellow tip
- Shorter tail than in Common/Arctic/ Roseate group, but longer than in Gull-billed
- Black legs (c.f. all except Gull-billed)
- Outer primaries make distinctive dark wedge ('sand wedge')

BLACK & WHITE-WINGED TERNS

INTRODUCTION Black Tern (*Chlidonias niger*) [L 24cm] localized summer visitor (April–October) to large freshwater marshes and lakes, requiring much emergent vegetation. More widespread in passage periods, when it is also found along coasts and on reservoirs. White-winged Tern (*Chlidonias leucopterus*) [L 22cm] rare passage visitor from the east, occurring in similar habitats.

The so-called marsh terns differ from the more familiar sea terns in feeding primarily on insects, not fish, in the breeding season, obtaining them by dipping elegantly down to snatch them from on or just above the water's surface (although sea terns can do this), and never diving in. They also have generally darker plumage than sea terns, are smaller and have much shorter, barely forked tails.

Juvenile

Black Tern (juvenile)

- Much darker plumage than in other terns
- Smaller than Common
- Smaller and rounder head than Common's
- Shorter tail than Common's; wings extend far beyond it, giving bird distinctive jizz
- Obvious dark patch ('thumb mark') at shoulder

Common Tern (juvenile/1st winter)

- Plenty of orange-red coloration at bill tip all summer and into autumn
- Fairly flat crowned
- Bright orange legs
- Whiter than juvenile Black (Arctic juvenile whiter still)

Black Tern (juvenile)

- Dark patch on shoulder
- Grey scaly saddle
- More black on head than in White-winged Black
- Grey rump

White-winged Tern (juvenile)

- Lacks dark shoulder-patch, so white breast-side gleams
- Darker brown saddle than in Black
- White rump
- Paler tail than Black's

Black Tern (juvenile)

- Longer bill than White-winged's
- Obvious spot on shoulder
- Extensive black around eye
- More pointed wings than White-winged's

White-winged Tern (juvenile)

- 'Like a Black Tern with Little Gull's shape' - more compact with a shorter bi
- Shorter tail than Black's
- Broader wings than Black's
- White breast-side

AUKS

INTRODUCTION Common Guillemot (*Uria aalge*) [L 42cm] widespread breeder on sheer cliffs and islands; winters on coasts. Brünnich's Guillemot (*Uria lomvia*) [L 42cm] Arctic equivalent, often breeding on even taller and sheerer cliffs. Winters at sea around Iceland and down to central Norway; vagrant further south. Razorbill (*Alca torda*) [L 40cm] widespread, breeding on wider ledges and also boulder piles; winters on coasts. Black Guillemot (*Cepphus grylle*) [L 35cm] breeds in Scandinavia, Iceland and the north of Britain and Ireland on rocky coasts and islands, and not on cliffs. Tends to winter around breeding areas. Little Auk (*Alle alle*) [L 20cm] Arctic nesting species (Iceland, Svalbard, Jan Mayen) on scree and boulder slopes. Disperses far out to sea and is regular in winter south to the British east coast. Atlantic Puffin (*Fratercula arctica*) [L 31cm] widespread on predator-free islands that have soil-covered cliffs for making burrows; winters at sea.

Adult summer

Common Guillemot

- Distinctly long-bodied profile on water; longer than Razorbill
- Often, but not always, shows streaks on flanks (particularly towards far north)
- Very blunt end, with very short tail

Brünnich's Guillemot

- Completely different bill shape from that of Common – much shorter and blunter
- White line along cutting edge of upper mandible
- No black stripes on flank
- Black of breast-sides meets on breast to make sharp tip (blunt in Common)

Razorbill

- More compact, smaller and shorter bodied than guillemots
- No streaks on flanks
- Long tail clearly protrudes well beyond wings, and often raised, giving bird distinctive 'banana shape'; tail looks as though it could prick a balloon

Common Guillemot (adult summer)

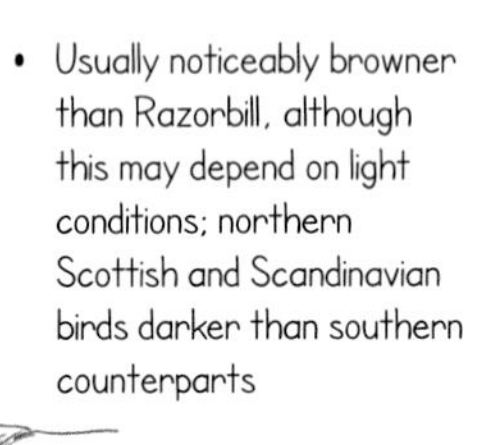

- Usually noticeably browner than Razorbill, although this may depend on light conditions; northern Scottish and Scandinavian birds darker than southern counterparts

'BRIDLED' FORM

- Small minority of adults have white eye-ring and whisker, and are known as 'bridled' guillemots - these become more common the further north you go
- Dagger-like bill, fairly long and pointed

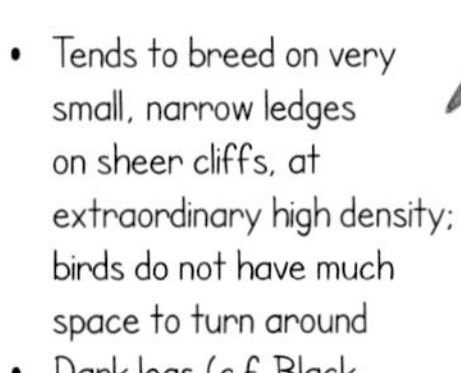

- Tends to breed on very small, narrow ledges on sheer cliffs, at extraordinary high density; birds do not have much space to turn around
- Dark legs (c.f. Black Guillemot, Atlantic Puffin)
- Thinner neck than Razorbill's

- Upright, penguin-like stance typical of auks

- Smudges on 'wing-pits' (rather clean white in Razorbill)
- Square tail

- Front profile distinctly pointed, even at long range
- Due to quite long rear end (protruding feet), in flight wings look placed slightly ahead of centre
- Hunchbacked profile distinctive (c.f. Razorbill)
- Tail short and feet stick out well behind in flight, unlike other auks'

Razorbill (adult summer)

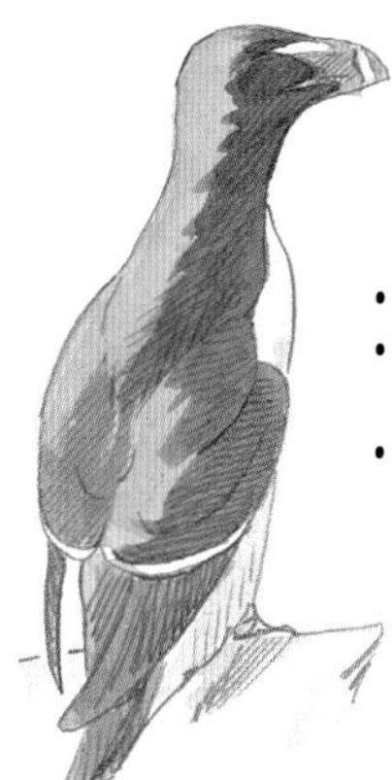

- Thick, powerful neck
- Very black above, not brown
- Flat crown

- Bill blunt ended; not at all pointed like Common Guillemot's
- Seen face-on, bill actually wafer thin (laterally flattened)
- White vertical line down bill
- White horizontal line from top of bill to eye

- Compact and thickset
- Long, pointed tail

- Thick neck
- Crown flat, not rounded
- Back of head angular
- Feet dull and dark (c.f. Black Guillemot, Atlantic Puffin)

- Looks balanced and wings appear centrally placed
- Bill often obvious, even at distance; at the very least, looks blunt at front
- Back straighter and much less hunched than in Common Guillemot
- Tail covers trailing legs, giving pointed rear end

Black Guillemot (adult summer)

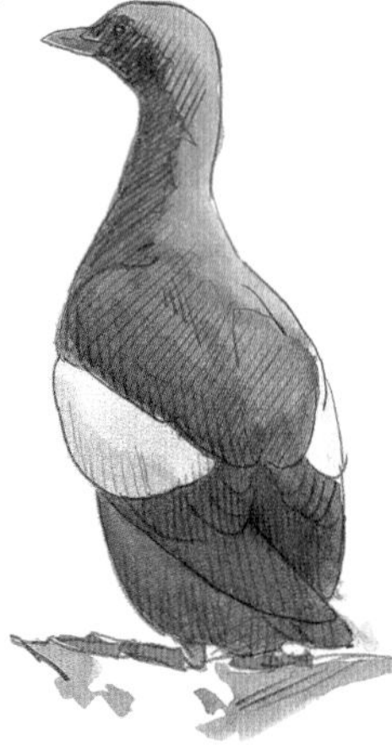

- Distinctly small head compared with bulbous body
- Small but pointed bill
- Very obvious red legs and feet

- Mainly smoky-black
- White oval pattern on coverts makes striking and unmistakable contrast to rest of dark plumage

- Thin, pointed bill
- Rounded head (has been described as 'dove-like')

- Takes off from water with long run, like other auks
- Red legs
- White wing ovals

- White wing-patch hard to miss, despite exceptionally fast wingbeats and low trajectory over sea
- Wings appear centrally placed
- Pointed tail
- Pot-bellied shape gives 'vent-heavy' look

- Smaller than Common Guillemot and Razorbill, but slightly larger than Atlantic Puffin
- More compact body than Common's
- Tends to look buoyant and sits high on water

1ST SUMMER
- Birds that fledged last season often attend colonies - distinguishable from adults by dark barring on oval wing-patches
- Often slightly browner than full adult

ADULT SUMMER AUKS FLYING AWAY

RAZORBILL
- Black upperparts
- More white on sides of rump than in Common Guillemot

COMMON GUILLEMOT
- Brown upperparts (darker in far north)
- Narrow white sides to rump
- Legs protrude

BLACK GUILLEMOT (JUST AIRBORNE)
- Clear white patches on upperwing
- Underwings white (as in Razorbill; lacks smudges of Common Guillemot)
- Red legs

Adult summer auks (continued)

Puffin (adult summer)

- Even more squat and clown-like than guillemots and Razorbill
- In breeding season, brilliant multicolored bill unmistakable
- Longer legs than other auks', and walks with ease among rocks or grass instead of clambering about awkwardly
- Legs bright orange

- Pale head can be best feature at great distance
- Front-heavy profile, with wings appearing set slightly backwards of centre
- No white trailing edges to wings (c.f. guillemots and Razorbill)
- No white on sides of rump (c.f. guillemots and Razorbill)

- Plump body
- Huge head
- Legs more centrally placed on body than in other auks, so can walk well and find it comfortable

JUVENILE

- Pale cheek-patch always visible, even on juvenile bird

JUVENILE

- Floats high, like a cork
- Short, squat body
- Big head
- Short tail just protrudes beyond wing-tip

- Completely dusky underwings (c.f. other auks)
- Flanks dark (but not streaked)
- Short tail, but legs do not protrude behind (c.f. guillemots)

Winter auks swimming

Little Auk

- Ear-coverts black
- Minute bill (eats plankton), much smaller than that of any other species
- White 'eyelid' (actually above eye)
- White 'V' marks on back (scapulars) (c.f. other auks)

Razorbill (both images)

- Heavy bill as obvious in winter as summer
- Thick, laterally flattened bill with blunt tip
- White band in middle of bill
- Diffuse white line from eye to top of bill
- Long, pin-shaped tail often held cocked up when swimming
- Clean white on underparts

Common Guillemot

- Black streak behind eye diagnostic
- Streaking on flanks variable but diagnostic
- White 'lobe' reaches up behind eye (c.f. Brünnich's)

Brünnich's Guillemot

- Stout, thick bill very obvious
- Black covers side of face, going down much further than in Common Guillemot and Razorbill
- Whitish gape-line
- Steeper forehead than that of Common and may show 'bump' on head
- White throat bordered by band across chest

Winter auks in flight

Black Guillemot

- Large white patches in wings present from breeding plumage
- White underwings bordered by black (unique)
- White head
- Barred back (unique)

Puffin

- Dark underwing (c.f. guillemots, Razorbill)
- No white trailing edge to wing (c.f. all except Black Guillemot)
- Dark flanks
- Grey head
- Bill not as large as in summer, but still substantial

Puffin

- Noticeably smaller than Razorbill and guillemots
- Large head
- Orange legs may be discernible

Little Auk

- White trailing edge (hard to see, but c.f. Atlantic Puffin, the only auk of even slightly comparable size)
- Dark underwings (as Atlantic Puffin, but not other auks)
- Whiter flanks than Atlantic Puffin's
- Only a suggestion of incomplete breast-band

Little Auk

- Bill barely visible
- Tiny - appears size of Starling
- Quite small, 'sunken' head
- Wings whirr; flight not always straight, tending to veer off at angle

Razorbill

- Thicker neck than in guillemots
- Whiter, cleaner underwings than in Common Guillemot
- Less hunchbacked than guillemots
- More black on face than Common Guillemot

Common Guillemot

- Relatively long body
- More of a collar than in Razorbill, but not much.
- Long, thin, pointed 'spiky looking' bill
- Relatively thin neck

Brünnich's Guillemot

- Much stouter bill than in Common
- Short neck gives stout appearance - shape often gives impression of rugby ball
- White armpits (dusky in Common)
- Half collar around neck
- Often flies with bill pointing slightly down (often slightly up in Razorbill, level in Common)

Common Guillemot

- Browner than Razorbill or Brünnich's Guillemot (but darker in far north)
- Legs trail behind (c.f. Razorbill)
- Less white on sides of rump than in Razorbill and, especially, Brünnich's Guillemot

Razorbill

- Tail longer than in guillemots, covering feet
- Broader white sides to rump than in Common Guillemot

PIGEONS & DOVES

INTRODUCTION Woodpigeon (*Columba palumbus*) [L 40cm] abundant widely in farmland, woodland edges, towns and cities. Stock Dove (*Columba oenas*) [L 30cm] widespread in woods and other places with mature trees, but generally avoids human habitation. Feral pigeon (Rock Dove) (*Columba livia*) [L 33cm] widespread resident over most of the region, particularly in urban areas, and on cliffs and mountains. Eurasian Collared Dove (*Streptopelia decaocto*) [L 32cm] suburban counterpart of feral pigeon, nesting in shrubs; widespread resident. European Turtle Dove (*Streptopelia turtur*) [L 28cm] summer visitor April–September to the Continent south of Denmark, including the southern half of Britain; mainly on arable land with hedgerows and in rural villages.

Feral pigeon is the domesticated descendant of Rock Dove, abundant in cities. It occurs in many colours (and even shapes), including greys, browns, white and black. If you see a flock of pigeon allsorts, they will invariably be feral pigeons.

Pigeons and doves in flight

Turtle Dove
- Dark outer wing
- Dark underwing contrasts strongly with white belly
- Distinctive fast flight, often swaying from one side to another

Collared Dove
- Whitish underwings do not contrast with belly
- Flight characterized by somewhat faltering, flicking wingbeats

Turtle Dove
- Unique tail pattern, with white 'string of pearls' at tip

Woodpigeon

- Landing Woodpigeon unmistakable, showing large white band in middle of each wing
- Dark outer wing

Stock Dove

- Ash-grey centre to wing
- Small double wing-bar close to body

Woodpigeon

- Striking white wing-band
- Long wings
- Flies on straight course with powerful, full wingbeats

Stock Dove

- Rather compact
- Grey centre and black outline to wing
- Dark underwing
- Fast flight, but wingbeats have more flickering style than those of Woodpigeon and feral pigeon (not as hesitant as Collared Dove's beats)
- Wings look shorter and stubbier than Woodpigeon's

Rock Dove

Note: this is the 'ancestral' form of Rock Dove. Feral pigeon occurs in flocks of mixed colours and patterns.

- Looks 'skinny' in flight
- Whitish underwing glints
- Two obvious black wing-bars
- Very swift flight with fast, unbroken wingbeats

Pigeons and doves perched

Turtle Dove
- Rich pinkish colour on breast
- Bluish head

Collared Dove
- Relatively small, with long tail
- Colour of creamy coffee
- Undertail pattern distinctive: black and white

Stock Dove
- Iridescent green patch; but no white on neck
- Much more compact than Woodpigeon
- Mainly subtle grey

Woodpigeon
- Big body and small head
- White patches on wings
- Rather long tail

Collared Dove
- Single black mark on neck makes half collar
- Larger and more portly than Turtle
- Rather uniform in colour

Turtle Dove
- Black-and-white neck-patch (zebra mark)
- Beautiful tortoiseshell pattern on back
- Red eye-ring

Woodpigeon
- Yellow eye
- Large white patch on side of neck

Stock Dove
- Black eye (c.f. Feral Pigeon red)
- Just iridescent green/ blue neck-patch
- Darker than Woodpigeon

SWIFTS

INTRODUCTION Common Swift (*Apus apus*) [L 17.5cm] common summer visitor to most of the region late April–August. Can be seen almost anywhere, but breeds in colonies on tall buildings. Pallid Swift (*Apus pallidus*) [L 17cm] just creeps into the region in southern France and Switzerland as a summer visitor April–October, on cliffs, mountains and buildings. Alpine Swift (*Apus melba*) [L 22cm] summer visitor April–September, mainly in mountains in southern France, Switzerland and northern Italy.

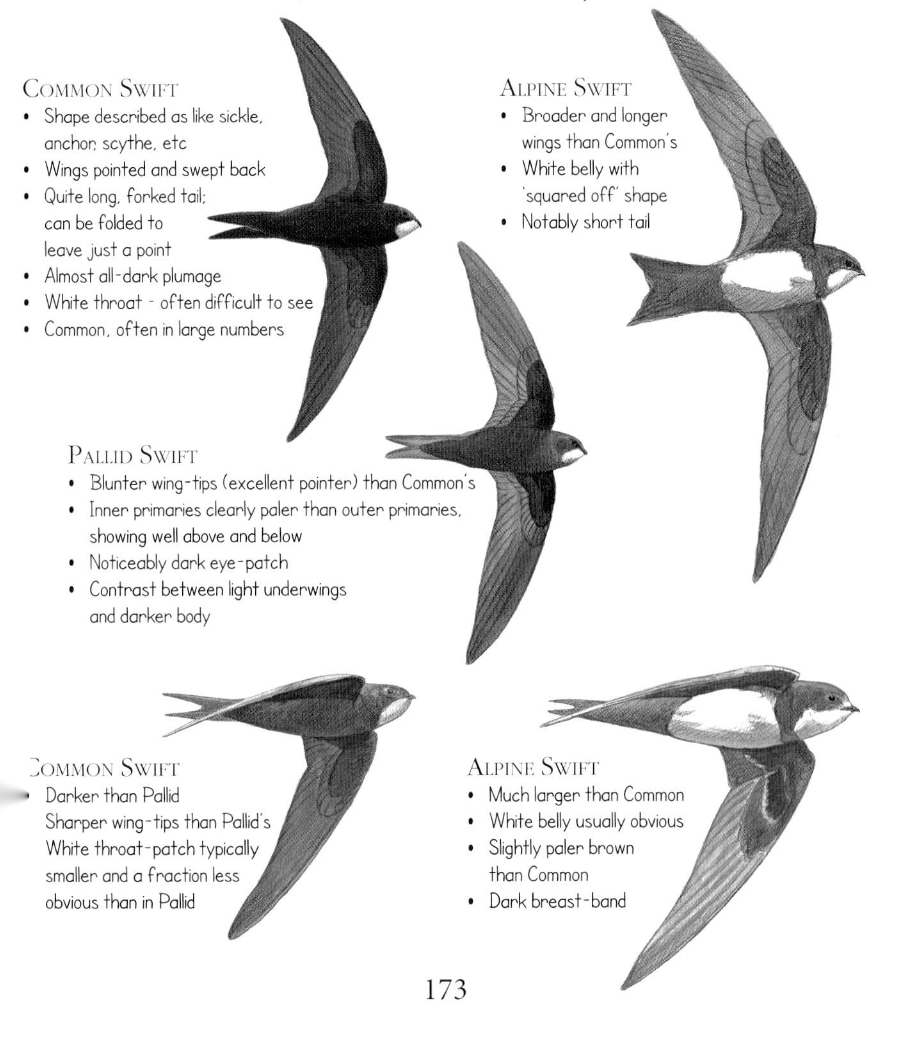

WOODPECKERS

INTRODUCTION Green Woodpecker (*Picus viridis*) [L 33cm] widespread resident in open woods, both deciduous and mixed, plus heaths and other places with ant-rich grassy swards and mature trees. Grey-headed Woodpecker (*Picus canus*) [L 28cm] scarce resident in southern Scandinavia, France and Germany, occurring in open woodland and parkland, especially near rivers and in hills. Great Spotted Woodpecker (*Dendrocopos major*) [L 25cm] common resident in all kinds of woodland throughout the region (not Ireland or Iceland); also parks and gardens. Lesser Spotted Woodpecker (*Dendrocopos minor*) [L 15.5cm] common and widespread but elusive resident in deciduous and mixed woods, also occurring in hedgerows and reedbeds. Middle Spotted Woodpecker (*Dendrocopos medius*) [L 21cm] uncommon and localized resident in mature oak woods on the Continent. All woodpeckers have distinctive undulating flight, in which a few power flaps (3–4) are alternated with closed wing descents.

GREEN WOODPECKER

FEMALE

- The most typical view – when flushed from ground, keeps most of body behind trunk, motionless, with an eye on observer
- Black around eye always obvious

FEMALE & YOUNG

- Black moustache indicates female
- Most 'angular' woodpecker, with long, chisel-like bill and triangular head
- Thinner, longer neck than in the other woodpeckers

MALE

- Sometimes sits still and fluffs feathers while perched
- Red moustache (c.f. female)

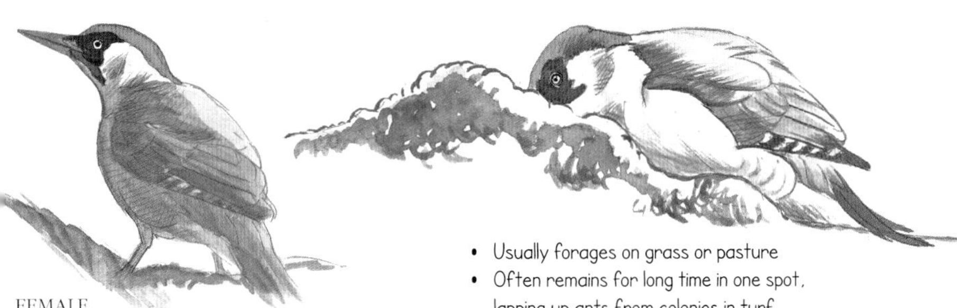

FEMALE
- Typical view of feeding bird looking up, alert
- When moving on ground, makes large hops

- Usually forages on grass or pasture
- Often remains for long time in one spot, lapping up ants from colonies in turf
- Makes heavy hacking movements to dig holes in turf

JUVENILE MALE
- Whitish scales on mantle
- Heavy streaks on face and neck
- Scaling on breast
- Red moustache indicates juvenile male; juvenile female has blackish streak

FEMALE
- Both 'green' species have yellow rump
- Extensive red cap

GREY-HEADED WOODPECKER

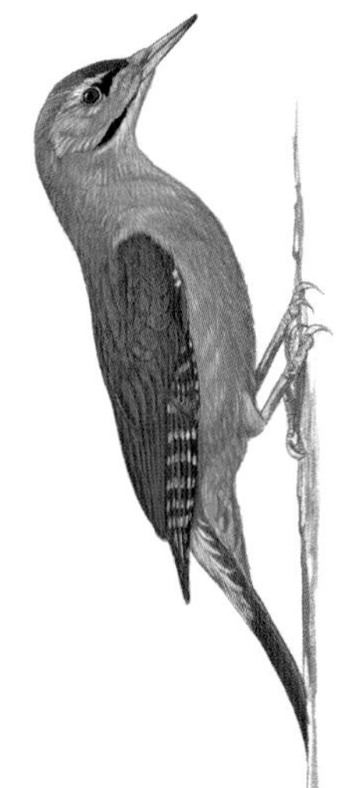

MALE
- Tail not clearly barred
- 'Snub-nosed', shorter bill

MALE
- Slightly smaller than Green, with distinctly smaller bill and shorter neck
- Ash-grey head and hint of grey on underparts
- Narrower black moustache than Green's
- Red confined to small patch on forehead (male only - female has grey crown)
- Red eye (looks dark at distance)

Woodpecker flight proportions

Green Woodpecker

- In contrast to the other woodpeckers, usually flies low to ground, not at treetop height
- When flying away, shows strongly yellow rump

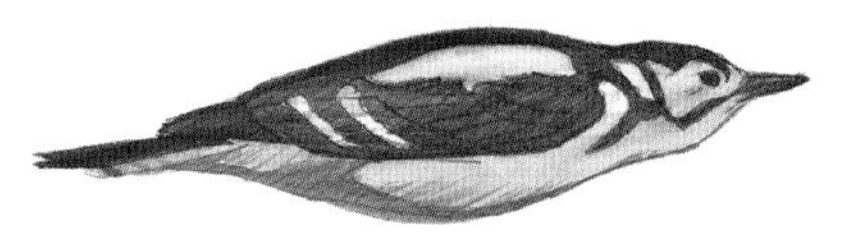

Great Spotted Woodpecker

- Considerably smaller than Green
- Large white slab on shoulder

Lesser Spotted Woodpecker

- Small - could be confused with large finch (never the case with Great Spotted)
- Smaller bill and head than Great Spotted's

Great Spotted Woodpecker

TYPICAL FLIGHT PATH

- Classic woodpecker flight pattern, extremely undulating: 3-7 flaps followed by brief closure of wings, then flaps again at bottom of slope

MALE, FROM BEHIND

- Distinctive large 'blobs' on back, like spilt paint (c.f. Lesser Spotted)
- Red nape patch indicates male

FEMALE

- This species tends to alight on larger trunks and limbs than Lesser Spotted, and less frequently on twigs in canopy

FEMALE

- Species often seen flying off from behind trunk
- Wings black with white bars

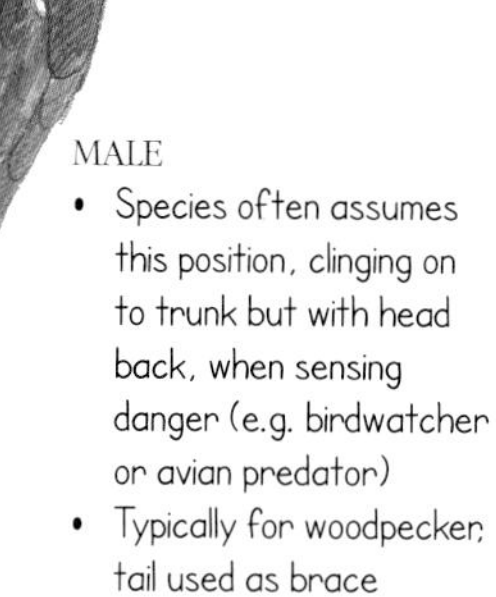

MALE

- Species often assumes this position, clinging on to trunk but with head back, when sensing danger (e.g. birdwatcher or avian predator)
- Typically for woodpecker, tail used as brace
- Moves upwards by hopping

FEMALE, FROM BEHIND

- Black nape indicates female

MALE

- Brilliant crimson on lower belly (c.f. Lesser Spotted, Middle Spotted)
- Breast white, unstreaked (c.f. Lesser Spotted, Middle Spotted)

'Pied' woodpeckers with red crowns

Lesser Spotted Woodpecker (male and juvenile)
- Lacks white shoulder-patches
- Barred white across back
- Tiny - size of a sparrow
- Very small bill

Middle Spotted Woodpecker (all plumages)
- Red crown lacks black lower border
- Moustachial stripe does not reach bill
- Strong streaking on pale buff breast
- Relatively short, weak bill
- Pink undertail-coverts
- Sexes similar

Great Spotted Woodpecker (juvenile, June-September)
- Red crown with black border; more red on crown of juvenile male
- Paler red undertail-coverts than in adults
- Slight barring on shoulder-patch (c.f. adult)

Lesser Spotted Woodpecker

TYPICAL FLIGHT PATH
- Less powerful flight than in Great Spotted, seemingly at lower speed
- Tendency to 'slip' from one part of tree to another using rather fluttering flight
- Small, short tailed - possible confusion with similar sized Nuthatch

FEMALE
- Species habitually hangs upside down, rather like a tit (Great Spotted does this too, but looks much more awkward when doing so)
- Usually more agile and fidgety than Great Spotted, moving about tree more rapidly

FEMALE, FROM BEHIND
• White lines along back, like rungs of ladder
• Rather short tail
FEMALE, DISTANT
• Looks short necked
• Short bill
• Look for it high up, towards top of tree
FEMALE
• White forehead and black crown
FEMALE
• Often clings to very narrow branches and twigs up high
• Streaks on breast
EMALE & BLUE TIT (FOR SIZE OMPARISON)
Much the smallest woodpecker, not very much larger than a tit
MALE
• Red crown indicates male (or juvenile)
• Barred back
• Heavily streaked on flanks

SHRIKES

INTRODUCTION Great Grey Shrike (*Lanius excubitor*) [L 24cm] breeds widely on bogs, heaths and tundra; summer visitor to northern parts; winters in similar habitats; scarce winter visitor to Britain. Iberian Grey Shrike (*Lanius meridionalis*) [L 24cm] equivalent species of southern France and Iberia, resident in open, bushy country with scattered trees. Lesser Grey Shrike (*Lanius minor*) [L 20cm] very rare breeding bird in parts of France and Germany May–September, and a rare migrant elsewhere; open, grassy country with scattered scrub. Red-backed Shrike (*Lanius collurio*) [L 17cm] local summer visitor to much of the region except the British Isles and northern Scandinavia, occurring in open country with scrub; scarce migrant to Britain. Woodchat Shrike (*Lanius senator*) [L 18cm] summer visitor to south, often in areas with more trees than is the case with other shrikes.

Grey shrikes

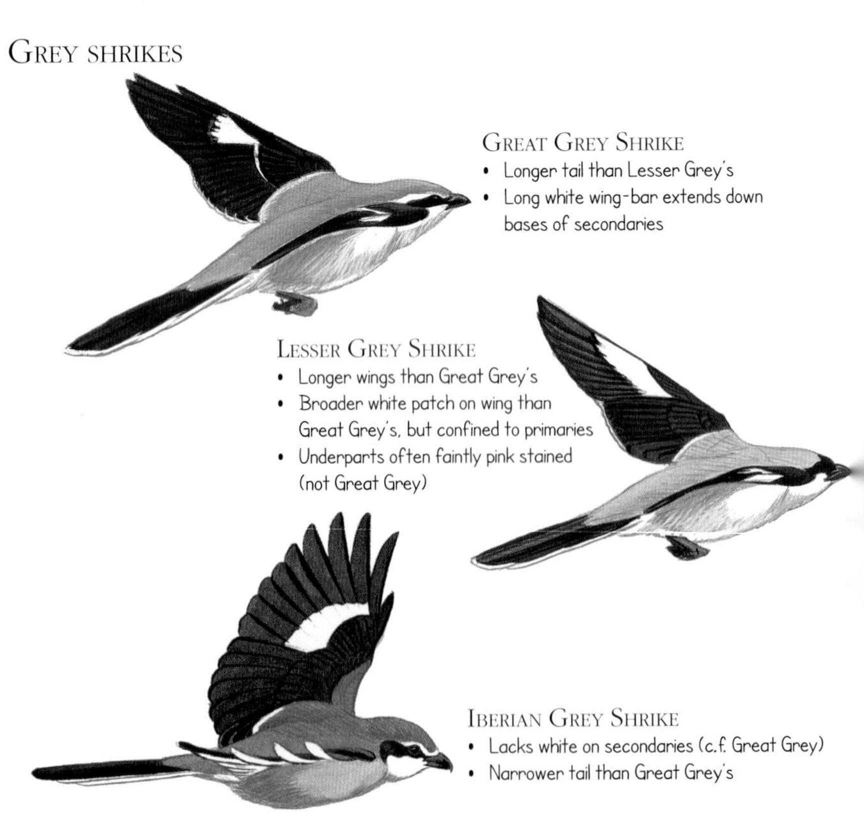

BERIAN GREY SHRIKE
Much darker above than Great Grey
More intense grey below than Great Grey (usually with pinkish tinge)
'Wavy' white supercilium to just behind eye
White on forehead above bill

LESSER GREY SHRIKE
- Square-ended tail
- Broad mask covers forehead
- Bold, rectangular white wing-patch
- Stubbier bill than in other shrikes

GREAT GREY SHRIKE
- Rather short wings (c.f. Lesser Grey)
- Broadest white edges to tail are near tip (near base in Lesser Grey)
- White scapulars
- Obvious white tips to tertials

LESSER GREY SHRIKE (JUVENILE/1ST WINTER)
- Scaly pattern to grey upperparts
- Scales on crown (not in juvenile Great Grey)
- Underparts plain, not scaled (c.f. Woodchat, Red-backed); faint barring
- Pale edges to flight feathers

MALL SHRIKES

ED-BACKED SHRIKE
ADULT MALE)
Stunningly smart bird, but smaller than you might expect - not much larger than a Starling
Chestnut-brown back
Pinkish underparts
Bold black mask with narrow white supercilium

WOODCHAT SHRIKE (JUVENILE)
- Whitish rump
- Ghost' of white shoulder-patch
- Scaly plumage and pale crown

Red-backed Shrike (1st winter)

- Grey crown and nape (c.f. juvenile)
- Back with some bars, but some grey as well

Red-backed Shrike (juvenile)

- Dark mask (helps to identify it as a shrike), but not as bold as adult male's
- Scaly below
- Rather warm, rusty colour above
- Crown warm brown, closely barred

Woodchat Shrike (juvenile)

- Pale crown with close barring
- Paler and more contrasting than Red-backed juvenile
- Scapulars whitish with dark crescents.
- Obvious whitish wing-bar made by greater coverts

Woodchat Shrike (juvenile)

- Paler and more variegated than Red-backed juvenile
- White patches at bases of primaries
- Whitish rump
- Mainly dark tail

Red-backed Shrike (juvenile)

- Rustier coloured than Woodchat juvenile and without obvious wing-patches
- Rusty-coloured tail

CROWS

INTRODUCTION Carrion Crow (*Corvus corone*) [L 47cm] ubiquitous resident over most of the region except northern Britain, Ireland and Scandinavia, where it is replaced by Hooded Crow (*Corvus cornix*) [L 48cm]. Rook (*Corvus frugilegus*) [L 45cm] common widespread resident of arable areas with scattered woods. Common Raven (*Corvus corax*) [L 60cm] resident in mountains and coastal regions and cliffs throughout the area, but scarcer in the lowlands (e.g. the Low Countries, south-east England and northern France). Western Jackdaw (*Corvus monedula*) [L 32cm] throughout in old woods, arable areas, towns and rocky places. Red-billed Chough (*Pyrrhocorax pyrrhocorax*) [L 39cm] uncommon resident on some cliffs and mountains in the British Isles and France; Alpine Chough (*Pyrrhocorax graculus*) [L 37cm] in high mountains, e.g. the Alps and Pyrenees.

Heads

Carrion Crow

- Flat crown
- Gentle slope down from crown, over bristles and on to bill
- Bristles at base of culmen (c.f. adult Rook)
- Culmen curves well down

Rook (juvenile, April–October)

- Peaked crown, as adult Rook
- Wedge of feathers on top of culmen base, as Carrion Crow, but bulges down on its lower edge
- Bill black (as Carrion Crow)
- Bill relatively narrow and pointed (as Rook)
- Culmen rather straight

Rook (adult)

- Peaked crown
- Bill and bill-base dirty white
- Bill thinner than Carrion Crow's, and more pointed
- Culmen less obviously curved than in Carrion Crow

Raven

- Long throat feathers give shaggy appearance, diagnostic to Raven (at times may also be visible in flight)

Adult crows

Raven

- Tends to inhabit wild places and less tolerant of humans and their trappings than either crows or Rooks

Hooded Crow

- Size and shape identical to Carrion Crow
- Dirty grey underparts, nape and mantle contrast sharply with black elsewhere on body and render species unmistakable

Raven

- Large head and enormous bill dominate profile
- Culmen more strongly arched downwards than in Carrion Crow
- Flat crown
- Fairly neat plumage and closely feathered legs give similar profile to Carrion Crow on ground (at distance, birds can look very similar)

Carrion Crow

- All black with minimal sheen
- Distinctly neat, well-groomed, tidy plumage
- Tightly feathered legs - 'wears leggings down to knee'
- Walks with sinister gait

Rook

- Slightly more obvious purplish-blue sheen to plumage than in Carrion Crow
- Somewhat shabby and poorly presented - as though it is wearing clothes that are a size too big
- 'Baggy shorts'
- Flattened breast
- Walks with rolling gait

Red-billed Chough
- Diagnostic long, downcurved scarlet bill (orange in juveniles in summer)
- Red legs
- Legs longer than those of other crows
- Glossier plumage than in other crows, with blue sheen, especially to wings

Alpine Chough
- Shorter legs than Red-billed's, but also red
- Yellow bill rather shorter than Red-billed's, and with less of downwards curve
- Tail projects beyond wings (c.f. Red-billed)
- Extreme habitat specialist: no lower than 1,500m and often much higher

---

Crows in flight

Red-billed Chough
- Expertly wheels in sky and makes spectacular, closed-winged tumbles towards ground, only to open its wings at the last moment and swoop back upwards
- Narrow head
- Broad-based wings with pronounced 'fingers'
- Short, rather square-ended tail
- Distinctive two-toned appearance below, with coverts darker than primaries and secondaries

Alpine Chough
- Longer tail than Red-billed's
- Narrower base to tail than in Red-billed
- Yellow bill
- Wings slightly 'pinched in' at bases (c.f. Red-billed) and more rounded

Jackdaw
- Smaller than Rook or Carrion Crow; about size of a pigeon, and similarly quick, fast wing-beats
- Eye pale (diagnostic); does not show in juvenile (June-September)
- Cheeks and nape distinctively thundercloud-grey
- Short bill easiest and most consistent feature separating Jackdaw from other crows
- Wings more monotone below than either chough's

Rook

- Hand (outside half) of wing narrower than base, giving more tapered outline
- Wing-tips with slightly longer fingers than Carrion Crow's
- Bases of wings 'pinched in'
- Tail slightly longer than Carrion Crow's, with more wedge-shaped tip when fanned

Carrion Crow

- Broad-based wings, not pinched in at base
- Slightly squarer wings than Rook's, without much tapering at wing-tip
- Very slightly shorter tail than Rook's, with squarer tip when fanned

Raven

- Head and neck project forwards from body to greater extent than in Rook and Carrion Crow
- Longer tail than in other crows, with outer feathers longer than inner ones, giving wedge-shaped tip
- Long, fairly narrow wings

Raven

- Huge bill and shaggy throat feathers still distinctive in flight
- Distinctly backswept wings
- Wings much longer than Rook and Carrion Crow's, and hand narrow and tapered

Flight style

Common Buzzard (juvenile)

- For size comparison

Raven

- Much bigger (and more intimidating) than Carrion Crow - as big as Common Buzzard
- Wings swept back
- Longer tail than crow's

Carrion Crow

- More compact than Raven, and flies with quicker wingbeats

Raven (tumbling)

- Frequently tumbles in flight, and may even fly upside down for short periods.

MARSH, WILLOW & COAL TITS

INTRODUCTION Coal Tit (*Periparus ater*) [L 10.5cm] common widespread resident in conifer woods and forests; also mixed woods and gardens, less often deciduous stands. Marsh Tit (*Poecile palustris*) [L 12cm] bird of broadleaved woodland, usually with good understory; occurs patchily in most of region. Willow Tit (*Poecile montana*) [L 12.5cm] occurs widely, including right up to the Arctic. In the north occurs in conifer stands where Marsh is absent; also deciduous woods, hedgerows and damp areas.

Marsh Tit

- Smart appearance
- Well proportioned, with even-sized head
- Usually smaller than Willow, with more neatly defined black bib
- Wing plain

Willow Tit

- Tends to be plumper and 'fluffier' looking than Marsh
- Looks bull necked - bigger head than Marsh's
- Slightly rounded tail
- Pale edges to secondaries usually produce pale mid-wing panel
- Larger head than Marsh's

Coal Tit

Main differences from Marsh & Willow

- White nape-patch (surprisingly obvious), giving 'badger-striped' head
- Two white wing-bars
- Different colour scheme than in Marsh and Willow: more olive-brown above and slightly dirtier below
- Large black throat-patch with uneven edges
- Not as smart as Marsh or Willow
- Proportionally larger head than Marsh and Willow's
- Proportionally shorter ('spiky') tail
- Thinner bill

Marsh Tit

- Glossy black cap
- Shorter cap than Willow's
- Cheeks white in front and faintly dusky behind, with contrast between two (less in Willow)

Willow Tit

- Black on cap not glossy - more matt black
- Black goes further down nape than in Marsh
- Larger extent of white on cheeks, going further back towards nape than in Marsh

Marsh Tit

- Small pale spot at base of upper mandible, just above cutting edge (not in Willow)

Willow Tit

- Sometimes pale cutting edge to mandibles, as in Marsh

Marsh Tit

- Longer tail and neater proportions than Coal's
- More white on head than in Coal
- Neat black cap and bib
- No wing-bars

Coal Tit

- Often seen from below, as it tends to feed high up in conifers
- Large amount of black on throat
- Two white wing-bars

LARKS

INTRODUCTION The larks are a group of ground-living birds with streaky brown plumage. They run along the grass or earth, and do not hop. Most have distinctive flight songs. Eurasian Skylark (*Alauda arvensis*) [L 17cm] widespread resident in many kinds of open country, including agricultural fields, grassland and saltmarshes; summer visitor to Scandinavia. Woodlark (*Lullula arborea*) [L 14.5cm] open areas with scattered trees, but uncommon in Britain and southern Scandinavia, and migratory in colder places. Shore Lark (*Eremophila alpestris*) [L 17.5cm] uncommon summer visitor to the Scandinavian tundra; winter visitor to beaches and dunes in Britain and the Low Countries. Crested Lark (*Galerida cristata*) [L 18cm] patchily distributed on the Continent in dry, open country, often by roadsides. Thekla Lark (*Galerida theklae*) [L 16cm] rare resident in extreme southern France, in rocky areas. Greater Short-toed Lark (*Calandrella brachydactyla*) [L 15cm] localized summer visitor to parts of France, most common in the south in dry, cultivated areas; elsewhere a rare migrant, mainly in autumn.

Larks vs pipits

Skylark

- Much larger and heavier than a pipit
- Thicker bill than in pipits
- Can show a crest (no pipit has crest)

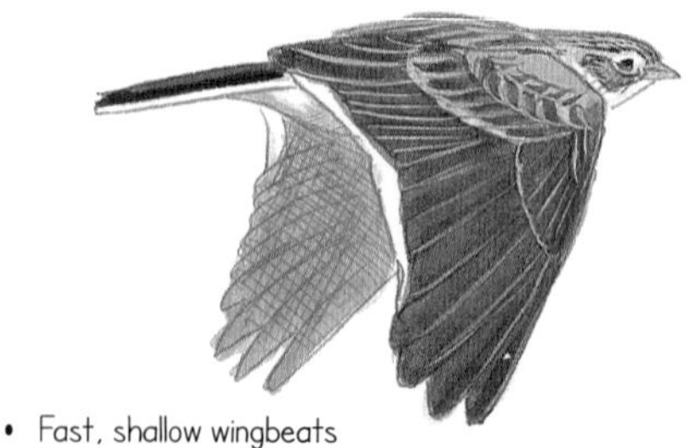

- Fast, shallow wingbeats
- Broader wings than a pipit's

Meadow Pipit

- Smaller than Skylark; often looks like tiny thrush
- Longer tailed and better proportioned than Skylark
- Neat pencil-lines on mantle

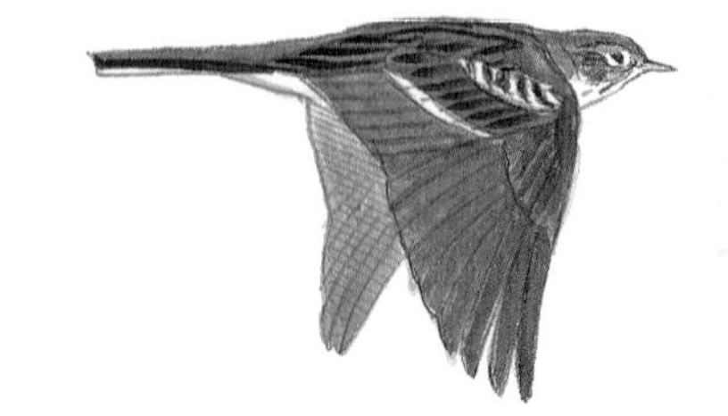

- Smaller and lighter than Skylark
- Lighter, more bouncy flight than Skylark's

Larks standing

Skylark (fresh autumn plumage)

- Longer tail than Woodlark's
- White outer-tail feathers (c.f. Woodlark, Crested Lark)
- Weak supercilium

Skylark

- Often seen on ground partly concealed in long grass
- Heavy but fairly long bill helps to distinguish it from other similar birds

Short-toed Lark (female, fresh autumn plumage)

- Smaller than other larks
- Thick bill rather finch-like
- Long tertials almost cover wing-tips
- Very little streaking below (c.f. others)

Short-toed Lark (male)

- Black spot/lines at sides of breast
- Median coverts dark centred, making wing-bar
- Characteristic jerky, hesitant walk

Woodlark (adult)

- Noticeably short tail
- Whiter on breast than Skylark
- Diagnostic pattern along edge of wing - narrow 'zebra-crossing' pattern (primary coverts and alula both dark based and pale tipped)
- Reddish-brown cheeks bordered by black at back edges

Woodlark

- Smaller and more compact than Skylark
- Very prominent, long supercilia that almost meet on back of head
- Tail has white tip, not white sides (as Skylark); tricky to see, while prominent white outer-tail feathers of Skylark usually very obvious
- Thinner and longer bill than Skylark's; slightly downcurved

Crested Lark

Differences from Skylark

- Much more obvious crest that points upwards, like a wisp of dried hair; almost always visible
- Outer-tail feathers chestnut, not white
- Shorter tail
- Longer bill with curved culmen

Thekla Lark

Differences from Crested

- Shorter bill with different shape; outer edges of mandibles curve towards each other
- Slightly shorter, more triangular crest; neater and often described as fan-like
- Heavier, bolder streaking on breast
- White eye-ring

Shore Lark (adult winter)

- White on breast and belly; no streaks, except faint ones towards flanks
- Black band on throat
- Unmistakable black and sulphur-yellow face pattern
- Distinct red-brown hue to shoulders (distinctive)

Shore Lark (adult winter)

- Almost unstreaked below
- Less heavily streaked on upperparts than Skylark or Woodlark
- Black legs

Larks in flight

Thekla Lark
- Cinnamon rump contrasts with back (in Crested rump browner)

Crested Lark
- Warm underwing colour
- No white trailing edge, as Thekla (c.f. Skylark)

Woodlark
- Wings lack white trailing edges characteristic of Skylark
- White wing-bar
- Blackish primary coverts

Skylark
- Most often seen from below, gaining height with rapid but shallow wingbeats, singing
- Medium-length tail

Skylark
- Looks heavy in flight - proportions quite similar to Starling's

Skylark (flying off)
- Rises without panic, often not very high if flushed
- Fast wingbeats often quickly turn into hover, or gentle undulations
- White trailing edge to wing distinctive

WOODLARK

- Very distinctive shape, with dumpy body and very short tail
- Rounder wings than Skylark's

SKYLARK

- White trailing edges to wings
- White outer-tail feathers
- Hindwing held straight; forewing angles back to blunt tips

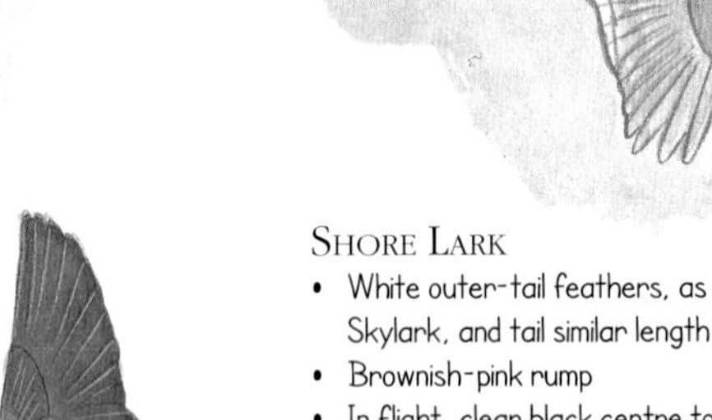

SHORE LARK

- White outer-tail feathers, as Skylark, and tail similar length
- Brownish-pink rump
- In flight, clear black centre to undertail (distinctive) contrasting with white underparts

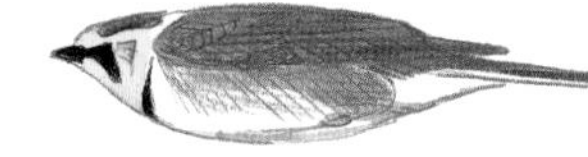

LARK HEADS

SHORE LARK (ADULT WINTER)

- Below crown, black stripe; rear feathers long, creating small horns that can project out above head
- Blackish bill (c.f. Skylark, Woodlark)
- Black face-mask

SHORE LARK (1ST WINTER)

- Slightly narrower black throat-band and mask than adult's
- Yellow slightly less intense than in adult
- Clean, red-brown nape

Skylark (winter)

- Generally looks open-faced
- Streaks on crown
- By late winter, streaks sharper and more contrasting

Skylark (spring, crest raised)

- Crest often raised
- Washed out (bleached) look in summer
- Nape may be almost whitish

Skylark (autumn)

- Warm brown coloration generally

Crested Lark (crest lowered)

- Longer bill than Skylark

Short-toed Lark
(male spring, bright individual)

- Diagnostic chestnut cap with black streaks
- White supercilium
- Black spot/lines at sides of breast
- Median coverts dark centred, making wing-bar
- Characteristic jerky, hesitant walk

SWALLOWS & MARTINS

INTRODUCTION Barn Swallow (*Hirundo rustica*) [L 18cm], Common House Martin (*Delichon urbicum*) [L 14.5cm] and Sand Martin (*Riparia riparia*) [L 13cm] summer visitors to the entire region, Sand Martin breeding mainly by rivers, lakes and sandpits, Barn Swallow in rural areas and House Martin in villages, towns and cities. Red-rumped Swallow (*Cecropis daurica*) [L 17cm] summer migrant to the very south of the region, in France, a rare overshooting stray further north. Eurasian Crag Martin (*Ptynoprogne rupestris*) [L 14.5cm] summer visitor to central and southern France, breeding on cliffs and mountains.

Upperparts

Barn Swallow

- Dark royal blue above without white rump
- Long wings
- Tail long with very narrow projecting streamers

House Martin

- Brilliant white rump contrasting with dark upperparts, even with poor view
- Shorter, less swept-back wings than Swift or Barn Swallow's
- Uninterrupted dark glossy-blue from crown to mantle (see Red-rumped Swallow)
- Tail well forked, but short

Sand Martin

- Entirely grey-brown
- No white rump
- Wings narrower than House Martin's, and shorter than Swift or Barn Swallow's

Underparts

Crag Martin

- Looks bigger and heavier than Sand Martin (which looks similar from above), and regularly glides
- Dark throat, thinly streaked, and no breast-band
- Short, broad tail with very shallow fork and pale 'windows' (visible also on uppertail)
- Notably dark undertail
- Very dark coverts contrast with rest of underwing

Common Swift (for comparison)

See also page 173

- Wings much narrower and more sharply pointed than in swallows or martins
- Wings all hand, arm remarkably short; swifts effectively fly with their fingers (c.f. swallows and martins)
- Bases of wings (where they join body) much narrower than in swallows and martins

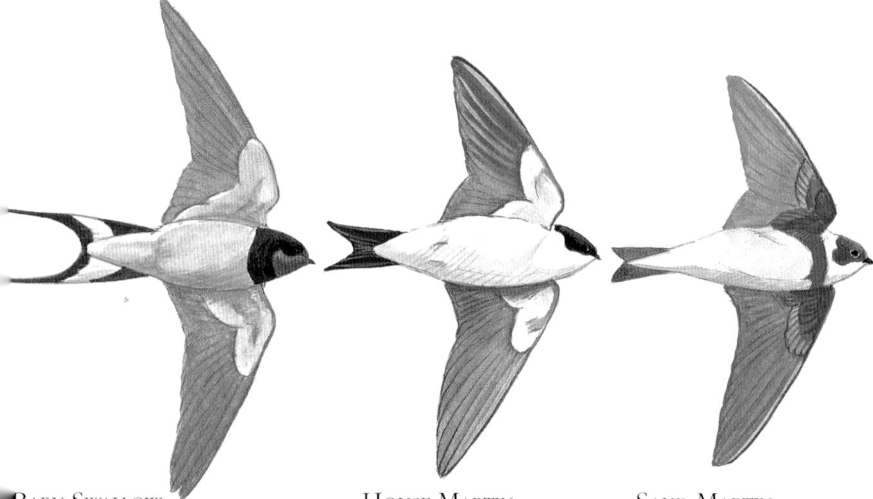

Barn Swallow

- Buff underparts contrast with dark chest-band
- Reddish throat
- Streamlined shape with long, forked tail

House Martin

- Black tail
- Gleaming white underparts
- Blue cap

Sand Martin

- Short tail
- Brown breast-band

Juveniles

Barn Swallow

- Signature long tail-streamers may be lacking, or much shorter than in adults
- Buff throat bordered by broad dark breast-band
- White markings in tail

House Martin

- Quite chunky in shape (c.f. Sand Martin) with bull neck
- Tail black (c.f. Sand Martin)
- Slightly grubbier underparts than in adults, but whiter than those of other common species here
- Juveniles distinguished from adults by dusky sides to neck (note superficial similarity to Sand)

Sand Martin

- Slim body compared with House's
- Weak fork to tail
- Complete breast-band (c.f. House Martin, Red-rumped Swallow)
- Wing-coverts darker than primaries, the reverse of House Martin and Barn Swallow (but as Crag Martin)
- Brown on coverts 'leaks' on to sides of breast (not in adult House Martin, so useful distinction in adult birds)

Red-rumped and Barn Swallows

Red-rumped Swallow

- At a distance can look like cross between Barn Swallow and House Martin, with tail of former and rump of latter
- Rump tends to be orangy (not red), and sometimes almost creamy towards tail
- Wings broad and House Martin-like, not as swept-back as Barn Swallow's; glides more often than Barn

Barn Swallow

- Dark royal blue above without white rump (c.f. House Martin)
- Distinctive white patches ('mirrors') on sides of tail distinguish it from martins and Red-rumped Swallow

Barn Swallow
- Creamy-white chest without streaks
- Royal blue head with orange-red throat
- Dark breast-band

Red-rumped Swallow
- Rusty-brown collar
- Mainly pale head (see Barn) allows eye to be visible
- Fine streaking on breast (stronger on adult, and diagnostic)
- Black tail lacking white markings

Habits and behaviour

House Martin
- Has to come down to ground to collect mud (as swallows)
- Note white-feathered feet

House Martin
- Builds nests under eaves of buildings
- Virtually always found in colonies (swallows often nest in single pairs)

Common Swift
- Partial to gathering in 'screaming parties', dashing around at rooftop height, skimming between buildings - swallows and martins do not do this

Barn Swallow
- Characteristically flies low, and often seen scooping drink from surfaces of ponds (the other species do this much less often)

Sand Martin
- Nests in colonies in sandbanks and other places where earth builds up, for instance at gravel pits

SYLVIA WARBLERS

INTRODUCTION Common Whitethroat (*Sylvia communis*) [L 14cm] widespread summer visitor April–September to scrubby places. Lesser Whitethroat (*Sylvia curruca*) [L 13cm] also widespread April–October, but missing from western France, northern Britain, Ireland and northern Scandinavia; in taller scrub than Common. Sardinian Warbler (*Sylvia melanocephala*) [L 13.5cm] resident southern France, in many scrubby habitats. Subalpine Warbler (*Sylvia cantillans*) [L 12.5cm] summer visitor to southern France, occurring on hillsides and mountainsides with scrub and trees. Spectacled Warbler (*Sylvia conspicillata*) [L 12.5cm] uncommon summer visitor to extreme southern France, where it occurs in low, often meagre bushy habitats and also scrub on saltings.

Common Whitethroat

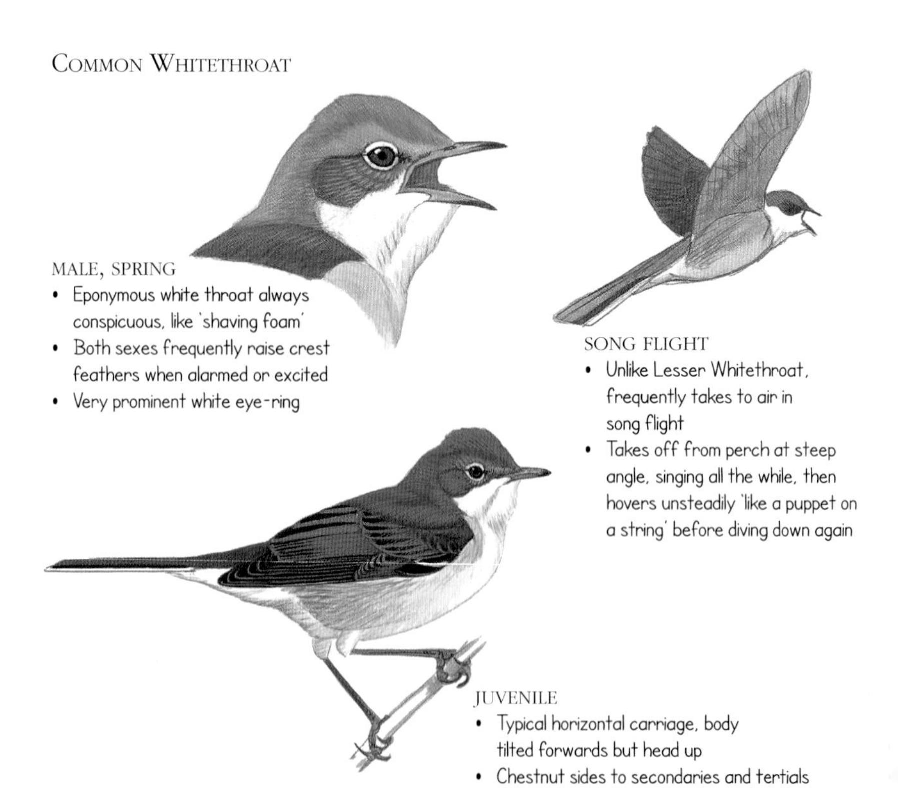

Sardinian Warbler

FEMALE
- Wing lacks chestnut tones (c.f. Common Whitethroat)
- Red orbital ring
- Darker underparts
- Long tail, short wings
- Dark grey head (c.f. female Subalpine and Spectacled Warblers)

MALE
- Black hood down to beneath eye
- Clean, contrasting white throat
- Red eye-ring very distinctive
- Slate-grey upperparts

Lesser Whitethroat

MALE, SPRING
- Typically looks sharply contrasting, unlike Common Whitethroat, dark upperparts setting off smart white underparts
- Darkness of ear-coverts varies individually
- Head more rounded than Common's, lacking any 'crested' look

MALE, SPRING
- Slightly smaller than Common Whitethroat
- Rounder head than Common
- Darker ear-coverts than Common
- Wings and back lack any rufous tones (c.f. Common)
- Sings from concealed perch (c.f. Common)

AUTUMN

Differences from spring bird
- Less strongly patterned; head tinged with brown
- Thin whitish supercilium

Differences from Common Whitethroat
- Legs black (diagnostic)
- Completely lacks any warm tone in wings (best distinction)

JUVENILE

Differences from adult
- More buff on underparts
- Pale fringes to wing-coverts

Spectacled Warbler

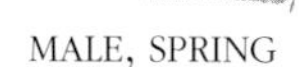

FEMALE

- Smaller and slighter than Common Whitethroat; diminutive, delicate-looking bird (might even recall *Phylloscopus* warbler)
- Much shorter primary projection
- Shares bright chestnut wings of Common Whitethroat, but they are plainer, with less black on tertials, which have tapering black centres

MALE, SPRING

- Black lores
- White eye-ring
- A little pinker on breast than Common Whitethroat

Subalpine Warbler

FEMALE

- Hint of white moustache
- Pinkish underparts
- White eye-ring outside reddish orbital ring
- Wings plain brown, without rusty tinge

MALE, SPRING

- Distinctive, with white moustache
- Warm brick-red colour on breast
- Dark grey hood and upperparts
- Red eye-ring

MALE, SPRING

- Subalpine usually sings from cover (Sardinian Warbler and Common Whitethroat typically from more exposed perch)
- White belly

FEMALE

- More pointed wings than in female Spectacled
- Plain wings without rusty colour rule out Common Whitethroat and Spectacled Warbler
- White moustache

GENERAL NOTES ON WARBLERS

INTRODUCTION Warblers are a large group of insect-eating songbirds that combine restlessness with generally retiring habits. Most are small and flit from perch to perch within dense foliage and make themselves difficult to see. This section covers the identification of major groups and some individual species.

Zitting Cisticola (*Cisticola juncidis*) [L 10.5cm] resident in overgrown, grassy places, agricultural fields and marshy areas in parts of France, especially the south; resident. Cetti's Warbler (*Cettia cetti*) [L 13.5cm] resident in thick vegetation within or beside lush wetlands, often along paths and ditches, in France, the Low Countries and southern Britain. Savi's Warbler (*Locustella luscinioides*) [L 14.5cm] widespread summer visitor to larger reedbeds except Scandinavia and Britain (where rare). Common Grasshopper Warbler (*Locustella naevia*) [L 13cm] widespread summer visitor March–September to low, scrubby habitats, including marshy ones. Garden Warbler (*Sylvia borin*) [L 14cm] widespread late April–August in woodland edges and tall scrub. Barred Warbler (*Sylvia nisoria*) [L 16.5cm] summer visitor to parts of Germany and extreme southern Scandinavia in tall scrub and hedgerows; otherwise scarce migrant.

WETLAND WARBLERS

ZITTING CISTICOLA

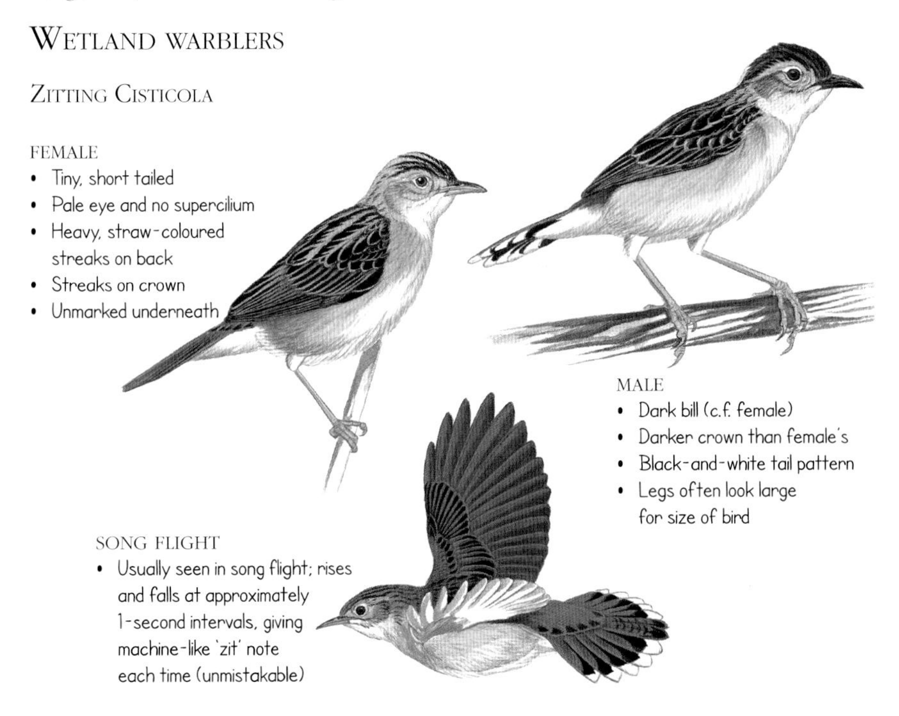

Sedge Warbler

- Sloping forehead and long bill typical of *Acrocephalus* warbler group
- Long, broad whitish supercilium (= Sedge)

Reed Warbler

- Long bill and sloping forehead, typical of *Acrocephalus* warbler group
- Long undertail-coverts also *Acrocephalus*
- Warm brown tones, especially on rump (c.f. *Hippolais* warblers, Savi's Warbler)

Cetti's Warbler

- Best identified by very broad, short, rounded tail, often held cocked
- Rich chestnut coloration
- Short wings
- Greyish on sides of head and breast
- Strong grey supercilium

Savi's Warbler

- Curved primaries and whitish edge to curved wing, typical of *Locustella*
- Similar colour to Reed Warbler, but typically *Locustella*, with long, rounded tail and heavy rear end
- Brown undertail-coverts (c.f. Reed Warbler)
- Rump same colour as back

Grasshopper Warbler

- Part of *Locustella* warbler group with more rounded crown and shorter bill than *Acrocephalus*
- A little more olive-brown than *Acrocephalus* and with faint eye-stripe at best
- Characteristic faint streaks on crown and sides of breast
- Overall appearance can recall Dunnock

Hippolais and *Phylloscopus* warblers

Icterine Warbler

- Featureless face compared with Willow and many other warblers - typical of *Hippolais* warblers (Icterine, Melodious, Olivaceous)
- No eye-stripe and plain lores
- Fairly peaked head similar to, for example, Reed Warbler and some *Sylvia* warblers, but not *Phylloscopus* warblers such as Chiffchaff and Willow
- Heavier, longer bill than in leaf warblers

Willow Warbler

- Small size, greenish-yellow plumage and prominent facial pattern typical of *Phylloscopus* warbler group
- Prominent eye-stripe and supercilium
- Fine, pointed bill

Icterine Warbler

More features of *Hippolais*:

- Squarer ended tails than in *Acrocephalus*
- Quite large, with long bills and sloping foreheads
- Broad-based bills

Willow Warbler

- Compared with *Hippolais*, much smaller, daintier and more spritely
- Sometimes makes flycatching sallies, or hovers (not *Hippolais*)
- Thin, flesh-coloured legs

Hippolais and *Acrocephalus* warblers

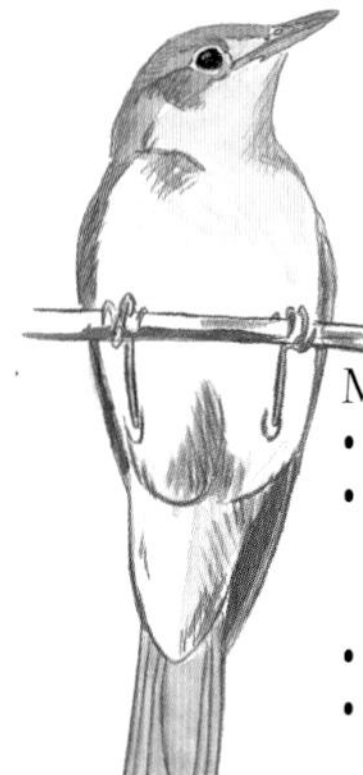

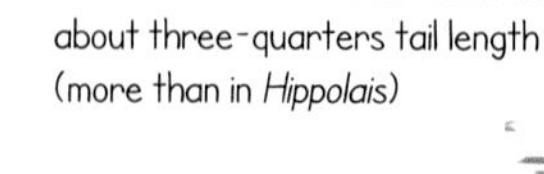

Reed Warbler (below)

- Undertail-coverts reach down to about three-quarters tail length (more than in *Hippolais*)

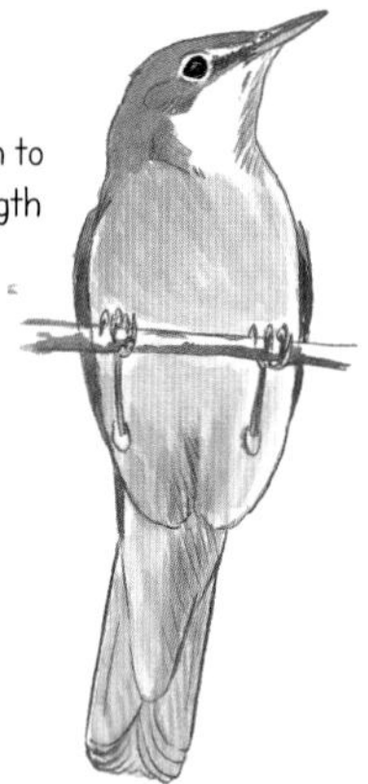

Melodious Warbler (below)

- Square-ended tail
- Yellowish wash to head and breast - always lacking in Reed and other *Acrocephalus* warblers
- Bland facial expression
- Hippolais often rather slow and clumsy for warblers; also quieter

Reed Warbler (structure)

- Undulating (banana-shaped) stretch from head to tail; *Hippolais* warblers straighter backed
- Typically clings to vertical stems - adapted to reedbeds
- Rounded tail

Garden and Barred Warblers

Garden Warbler (autumn)

- An atypical *Sylvia* warbler with unwarbler-like, short, thick bill
- Famously featureless species with plain upperparts and underparts
- Dark eye stands out clearly to give blank expression
- Grey on sides of neck
- Slate-grey legs

Barred Warbler (juvenile)

- Larger and bulkier, with longer bill than Garden's
- White outer-tail feathers
- Scaling on uppertail-coverts and especially undertail-coverts
- Flight feathers with obvious pale fringes

WILLOW WARBLER, CHIFFCHAFF & WOOD WARBLER

INTRODUCTION Willow Warbler (*Phylloscopus trochilus*) [L 12cm] widespread summer visitor March–September to woodland edge, especially birch; common migrant. Common Chiffchaff (*Phylloscopus collybita*) [L 11cm] widely distributed summer migrant March–October, or resident in mature deciduous woodland, scrub in winter. Wood Warbler (*Phylloscopus sibilatrix*) [L 12cm] widespread summer visitor April–August to mature oak and beech woods, and some mixed woods. Western Bonelli's Warbler (*Phylloscopus bonelli*) [L 11cm] summer visitor to France, Germany and the Alps, usually woodland in hills. Iberian Chiffchaff (*Phylloscopus ibericus*) [L 11.5cm] summer migrant to south-west France only, in mixed and other woodland; rare elsewhere. Yellow-browed Warbler (*Phylloscopus inornatus*) [L 10cm] rare migrant to much of the region, especially in late September–November; chiefly coastal.

Willow Warbler races

trochilus most of Europe.
acredula northern Scandinavia.

Common Chiffchaff races

collybita most of region.
abietinus central and northern Scandinavia.
tristis 'Siberian' Chiffchaff rare but regular migrant.

Adult spring

Willow Warbler

- Yellower than Chiffchaff, especially on face and upper breast
- Pale, flesh-coloured legs
- Longer supercilium than Chiffchaff's
- Yellower ear-coverts than Chiffchaff's give more 'open' face

Chiffchaff

- Slight darker green above than Willow, and thus more contrasting upperparts/underparts
- A trifle plumper and less elongated than Willow Warbler Black legs
- Dark ear-coverts

Wood Warbler

- Overall a brightly coloured bird, much more showy than Willow Warbler and Chiffchaff
- Chesty, even when not singing
- Wing more strongly patterned than in Willow Warbler and Chiffchaff
- Tertials noticeably dark with pale fringes

CHIFFCHAFF
- Darker ear-coverts than Willow's
- Eye-ring more obvious than Willow's because of darker ear-coverts

WILLOW WARBLER
- Slightly yellower supercilium than in Chiffchaff
- Face looks overall a little more washed through with yellow than Chiffchaff's
- Slightly finer bill than Chiffchaff's

WOOD WARBLER
- Much the brightest of the three, with brilliant, buttery-yellow face and, especially, supercilium

---

CHIFFCHAFF

- Black legs
- Shorter wings than Willow and Wood Warblers'

- Short wing-tips: primaries do not jut out far from end of tertials; only to about half the length of the tertials themselves
- Cheeks (ear-coverts) uniformly coloured, not mottled
- Dark eye-stripe splits white eye-ring

- Short, rather dark bill
- Quite dark, olive-green above
- Rather plain wings

TAIL-DIPPING
- Habitually 'wags' or flicks tail down while feeding; other species wag tail only periodically

WING-FLICKING
- Restless bird, flicking wings frequently

Wood Warbler

- Colour of upperparts much more intense green than in Willow Warbler and Chiffchaff
- Brilliant primrose-yellow throat and upper breast
- Very long, bright, intense yellow supercilium
- Contrastingly white on belly
- Strongly marked dark eye-stripe
- Ear-coverts bright yellow (without trace of pale eye-ring visible)
- Legs pale pink (variable)

- Very long wings - even longer than Willow Warbler's
- Noticeably short tail
- Broad shoulders give distinctive triangular shape
- Strong yellow on throat and upper breast

Willow Warbler

- Whiter on belly than Chiffchaff, but not as gleaming white as Wood Warbler
- Usually some delicate pale yellow on upper breast (Chiffchaff is darker buff)
- Longer bill than Chiffchaff's, usually with some pink at base of lower mandible

- More boldly patterned face, with more prominent supercilium, than Chiffchaff's ('sharper expression')
- Flatter crown than Chiffchaff's
- Altogether sleeker and less podgy than Chiffchaff
- Longer wings give it more darting, dashing flight than Chiffchaff's

- White offers more contrast to plumage (upperparts vs underparts) than in Chiffchaff
- Greyer and slightly paler above than Chiffchaff
- Ear-coverts may look blotched (uniform in Chiffchaff)
- Darker and more obvious eye-stripe than in Chiffchaff

- Ear-coverts pale near eye, so white eye-ring hard to see
- Longer wings than Chiffchaff's; primary projection - wing-tips project beyond tertials to length about the same as that of tertials themselves

Autumn leaf warblers

Willow Warbler (juvenile)

- Juvenile Willows (and Chiffchaffs) yellower than spring adults
- Yellower on supercilium than adult (and equivalent Chiffchaff)

Willow Warbler (juvenile/1st winter)

- Underparts quite strongly yellow
- Strong and well-defined yellow supercilium
- Often orange base to bill (c.f. Chiffchaff)

Chiffchaff (juvenile)

- Much clearer white eye-ring (especially below eye) than Willow Warbler's

Chiffchaff (juvenile/1st winter)

- Much less yellow on underparts than Willow Warbler
- Pale supercilium less obvious than Willow's
- Dark, small bill

Chiffchaff (typical adult)

- Supercilium somewhat indistinct; yellow wash (c.f. Siberian)
- Olive-green above, more colourful than northern or Siberian Chiffchaffs'
- Yellowish on underparts
- Dark eye-stripe splits pale eye-ring

Wood Warbler

- Brilliant yellow supercilium
- Strong yellow on breast
- Yellow ear-coverts
- Pale edges to tertials (c.f. Willow)
- Bright green edges to most wing feathers

Other subspecies and species

Willow Warbler (*acredula*)

- Long, pale, well-defined supercilium typical of Willow, as opposed to Chiffchaff
- Northern race Willows often greyer above than southern ones, looking very dull and unlike typical yellowish birds
- Paler below than usual Willow

Yellow-browed Warbler (1st winter)

- Very small - almost Goldcrest sized - and extremely active
- Huge, long whitish supercilium
- Two very obvious white wing-bars
- White spots on tertials

'Siberian' Chiffchaff (*tristis*)

- Colourless version of Chiffchaff, distinctly greyish on crown, nape and mantle
- Cold whitish underparts
- Well-marked supercilium has distinct buff tones
- Warm buff tinge to ear-coverts and neck
- Upper eye-ring often hidden within supercilium; lower eye-ring not as obvious as in typical Chiffchaff

Western Bonelli's Warbler (1st winter)

- Dull greyish-green back with contrasting yellowish rump
- Clean silvery-white underparts
- Rather plain-faced, with weak supercilium; dark eye stands out strongly
- Tertials obviously white edged
- Dark tail with obvious green edges (like Wood Warbler)
- Bright yellow-green wing-panel
- Surprisingly thick, pale bill and large head; rather tit-like

Chiffchaff (*abietinus*)

- Dull version of 'normal' Chiffchaff, but retains some greenish coloration
- Greyer-brown than normal Chiffchaff, with paler, less buff-coloured underparts - but not as clearly grey as 'Siberian'
- Supercilium tends to be stronger than normal Chiffchaff's

Iberian Chiffchaff

- Supercilium often brighter than Common Chiffchaff's, yellow in front of eye
- Slightly whiter on underparts than Chiffchaff
- Greener upperparts than Chiffchaff's, especially on rump
- Long, fine bill, with yellow lower mandible
- Slightly paler legs than Chiffchaff's

ICTERINE & MELODIOUS WARBLERS

INTRODUCTION Icterine Warbler (*Hippolais icterina*) [L 13cm] summer visitor to Scandinavia and continental Europe west to northern France, in open broadleaved woodland, often with tall trees. Melodious Warbler (*Hippolais polyglotta*) [L 12.5cm] summer visitor to south-west Europe, including Belgium and most of France, plus Switzerland and Italy, occurring in more scrubby habitats than Icterine.

Shape and structure

Icterine Warbler

- Heavy, broad-based bill with distinct pinkish-orange coloration on lower mandible
- Pale lores, so quite 'blank' expression
- Long, very pointed wings; wing-tips project well beyond tertials
- Often a clear wing-panel caused by pale edges to secondaries and tertials
- Greyish legs (c.f. Melodious Warbler)

Melodious Warbler

- Bill similar in shape and colour to Icterine's, but distinctly shorter - said to give bird a more 'friendly' expression
- Shorter wings than Icterine's (crucial); primaries project beyond tertials only to distance of less than half length of tertials, rather than almost whole length as in Icterine
- Shares Icterine's lack of distinct supercilium or eye-stripe, so looks featureless
- Wing tends to be plain, lacking any panel

Icterine Warbler

- Quite a big, powerful warbler, and dynamic, with swooping flight.
- Weak facial expression
- Very pale, especially below

Melodious Warbler

- Typically looks a bit 'portly'; quite lazy and inactive for a warbler, and skulking
- Olive-green upperparts
- Tends to look rather buffy-yellow below, not as pale as Icterine; slightly stronger wash of yellowish on throat and head
- Large, broad-based, orangy bill, rather blunt tipped

Juveniles

Melodious Warbler

- Some juveniles show stronger wing-panel than adults, more similar to Icterine's
- Slightly darker primaries than Icterine's
- Pale legs
- Upperparts fractionally less greyish than in Icterine

Icterine Warbler

- Browner than adult
- Usually more obvious secondary panel than in Melodious, but difference not as clear as in adults
- Greater wing-coverts show pale fringes

REED & RELATED WARBLERS

INTRODUCTION Eurasian Reed Warbler (*Acrocephalus scirpaceus*) [L 13.5cm] widespread summer visitor April–September to reedbeds; other habitats on migration. Marsh Warbler (*Acrocephalus palustris*) [L 14cm] late-arriving summer visitor May–September to southern Scandinavia, Germany, the Low Countries and France; rare in Britain. It is not a reedbed bird, instead inhabiting herbage by the waterside and sometimes in drier habitats. Sedge Warbler (*Acrocephalus schoenobaenus*) [L 12.5cm] widespread summer visitor April–October to scrub by or in wetlands; any scrub on migration. Aquatic Warbler (*Acrocephalus paludicola*) [L 12.5cm] migrant from breeding areas in eastern Europe usually seen in autumn; regular on French coast August–September, and rare elsewhere.

Reed Warbler (general features)

- Classic head/bill shape of *Acrocephalus*, with long bill and sloping forehead giving 'sharp-headed' appearance
- Similar to colour of reed stems - mid-brown above and buff below
- Warmest colour on rump

- Well adapted to clinging to vertical stems (strong legs); moves around restlessly, often hopping sideways from stem to stem
- Supercilium most obvious in front of eye (c.f. Sedge, Aquatic)

Spring adults

Reed Warbler

- Quite warm-brown in colour above, particularly on rump, which contrasts with mantle
- Fairly warm buff below
- White throat contrasts with buff breast (c.f. Marsh)
- Darker legs than Marsh (usually)
- Eye-ring brighter and more obvious than supercilium (about the same in Marsh)

Marsh Warbler

- When singing, holds mouth open wider than Reed Warbler, so orange of gape more obvious
- Often sings from dead stems of herbaceous plants

Marsh Warbler

- Slightly colder in colour above than Reed, olive-brown as opposed to brown; no difference on rump
- Washed subtle yellowish on underparts
- Primary tips clearly pale tipped, and tips well spaced
- Supercilium less distinct than in Reed

Sedge Warbler

- Broad, bold whitish supercilium
- Black eye-stripe
- Dark-streaked crown
- Streaks on back

Sedge Warbler (song flight)

- In contrast to Reed Warbler (and Marsh), often interrupts bouts of perch-singing with brief, low and somewhat feeble song flights

Juveniles

Reed Warbler

- More sharply pointed bill than Marsh's
- Shade warmer brown above than Marsh, especially on rump
- Often slightly warmer buff below than Marsh
- Primary tips less evenly spaced (some more bunched together) than in Marsh

Marsh Warbler

- Shorter bill than Reed's
- Paler legs than Reed's (especially claws)
- Primary tips more evenly spaced than in Reed, and brighter
- Slightly longer primary projection than in Reed

Sedge Warbler

- Dark lores (c.f. Aquatic Warbler)
- A few small streaks at edge of wing (not in adult, or Aquatic Warbler)
- Warm chestnut-coloured rump
- Browner legs than Aquatic's

Aquatic Warbler

- Pale buff centre to crown with dark lateral stripes either side
- Pale 'tram-lines' down mantle, contrasting with dark back
- Paler ground colour than in Sedge (yellower)
- Contrasting wings with very dark patches, but also broad pale fringes to many feathers (e.g. tertials)
- Pointed tail feathers

STARLINGS

INTRODUCTION Common Starling (*Sturnus vulgaris*) [L 21cm] widespread and ubiquitous throughout the region in many habitats; mainly summer visitor to Scandinavia. Spotless Starling (*Sturnus unicolor*) [L 21cm] rare (but increasing) breeding bird in the south of France. Rose-coloured Starling (*Pastor roseus*) [L 21cm] summer visitor June onwards, from the east, with some individuals, principally juveniles, remaining in autumn.

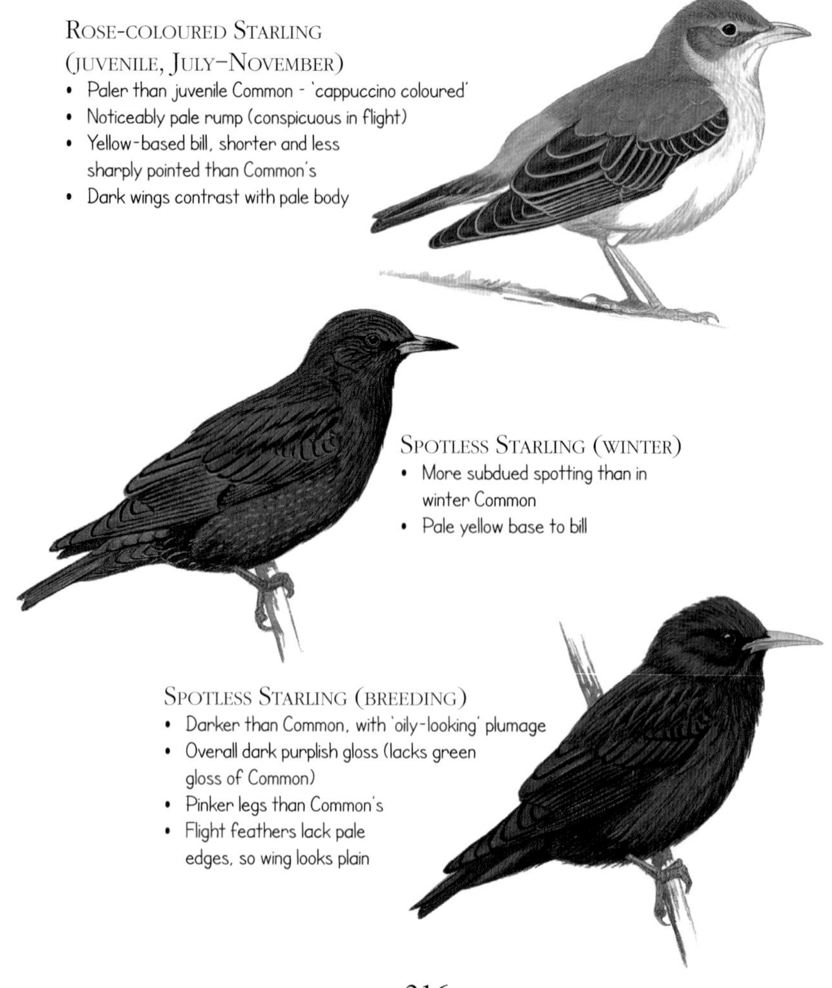

Ageing and sexing Common Starling
juvenile
• Dull, dirty-brown, featureless
• Shape indicates species: short tail, pointed wings, pointed bill
• Poorly defined streaks on underside
late juvenile, July–August
• Dark bill
• Head and neck feathers last to change
adult winter, October–February; (first-winter much the same)
• Dark bill
• Dense, star-like spots on body plumage (give bird its name)
• White spots on underside; some yellowish spots on upperside
• Pale edges to flight feathers
adult female breeding
• Yellow bill with pinkish base
• Some spots retained on body; females generally with larger spots on plumage than males
adult male breeding
• Yellow bill has bluish base (blue for male, pink for female)
• Most spots have worn off, giving all-over smart metallic sheen
• Glossier than female

SPOTTED THRUSHES

INTRODUCTION Mistle Thrush (*Turdus viscivorus*) [L 27.5cm] widespread resident where open grassy areas abut woodland; summer visitor to Scandinavia. Song Thrush (*Turdus philomelos*) [L 21cm] common resident in woods with good undergrowth, plus gardens, parks and other bushy places; summer visitor to parts of Germany, and Scandinavia. Redwing (*Turdus iliacus*) [L 21cm] breeds in Scandinavia, Iceland in birch woods and scrub; common passage migrant and winter visitor elsewhere in fields, hedgerows and woods.

Song Thrush

Mistle Thrush

ADULT, TYPICAL POSE

- Mistle Thrush much larger than Blackbird
- Small head
- Pot bellied

ADULT

- Rather blank expression
- Larger bill than Song Thrush's
- Dark edges to ear-coverts

ADULT

- Longer tail than Song Thrush's
- Bulging 'beer-gut' tummy and small head give uneven appearance
- Tail with white sides at tip
- Pale wing-panel on secondaries
- Plain grey-brown mantle
- Secondaries and tertials with white fringes
- Better developed breast-spots than on youngster

ADULT

- Stands tall and confident, with more obvious upright stance than Song Thrush
- Rounded spots on clear white background
- Spots often coalesce into smudge on breast-sides
- Wings often held drooped

JUVENILE

- Large white 'teardrop' marks on mantle and scapulars
- A few white streaks on crown
- Flight-feather fringes warm buff

Mistle Thrush

- Different flight style from Song Thrush and Blackbird's; low and deeply undulating (may recall a Green Woodpecker)
- White underwings (reddish-brown in Song Thrush)
- White sides and tips to tail (not in Song Thrush)

Redwing and Song Thrush

Redwing

- Broad buff supercilium key feature
- Chestnut-red flanks - colour 'leaks' down from underwings
- Streaks, not spots on breast
- Underparts whiter than Song Thrush's

Song Thrush

- Lacks pale supercilium
- Flank and breast buff

Female and juvenile Blackbird

Adult female

- Whitish throat, heavily streaked
- Streaks on upper breast set against much darker ground colour than in Song or Mistle Thrush
- Yellowish bill

Juvenile

- On some juveniles, streaks give strong impression of Song Thrush
- More rufous ground colour on breast than in female Blackbird, but colour still much darker than pale buff breast of Song Thrush
- Dark bill (c.f. female)

BLACKBIRD & RING OUZEL

INTRODUCTION Common Blackbird (*Turdus merula*) [L 26cm] ubiquitous in habitats including lowland woods, forests, gardens, hedgerows and agricultural areas. Ring Ouzel (*Turdus torquatus*) [L 25.5cm] summer visitor late March–October to hillsides and moorland, plus coniferous forest, usually above 250m ; widespread on migration. Highly distinctive race with heavy body scaling occurs in Central European mountains (not included here).

Ring Ouzel (male)

- Slimmer and less bulky than Blackbird
- Flatter back than Blackbird's
- Shape of white band distinctive and rarely matched precisely by albino-type Blackbird
- Yellow bill with black tip

Ring Ouzel (female summer)

- Usually less obvious and bright white crescent (some white feathers dark tipped) - but some females hard to tell apart from males
- Usually browner than male
- Usually more scaly on underparts than male
- Bill usually paler than male's

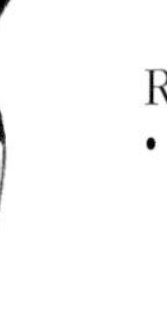

Ring Ouzel (female)

- Flight feathers have silvery edges - often most reliable indicator of Ring Ouzel rather than Blackbird

Blackbird (male, partially albino)

- Some males show variable white patches on breast, which can look similar to pattern of Ring Ouzel
- Yellow bill
- Yellow eye-ring
- Lacks pale scaling on plumage

Ring Ouzel (male summer)

- Clear black below bright white gorget (not usually in female)
- Yellow bill with darker tip; bill slightly thicker than Blackbird's
- Lacks male Blackbird's yellow eye-ring

Ring Ouzel (1st-winter male)

- Similar to adult female Ring Ouzel
- Dark bill (as first-winter male Blackbird)
- White band less obvious than in adult, but always present
- Scaly plumage (caused by pale tips to feathers)

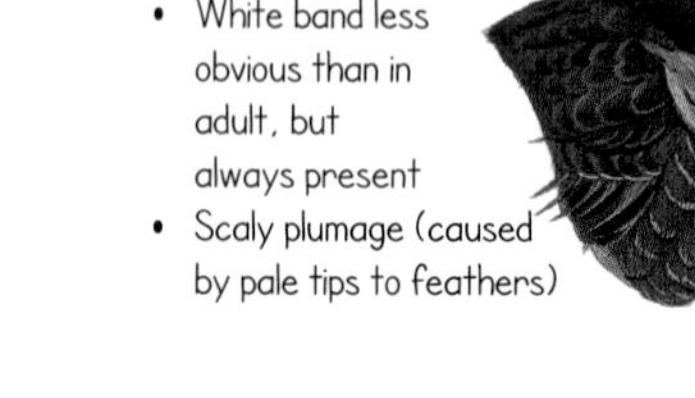

COMMON & BLACK REDSTARTS

INTRODUCTION Common Redstart (*Phoenicurus phoenicurus*) [L 14cm] widespread summer visitor April–October in open woods, both deciduous and mixed; common migrant in scrub. Black Redstart (*Phoenicurus ochruros*) [L 14cm] common continental species (scarce resident in Britain) in rocky places and on buildings of all kinds, also gardens; mainly summer visitor, but some winter in south. Common Nightingale (*Luscinia megarhynchos*) [L 16cm] widespread summer visitor April–August, not northern Britain, Ireland or Scandinavia; breeds in scrubby and bushy areas. Bluethroat (*Luscinia svecica*) [L 13.5cm] localized summer visitor to Scandinavia, the Low Countries and parts of France and Germany, breeding in generally rich, damp, scrubby habitats; rare migrant in Britain.

COMMON REDSTART (FEMALE)
- Redstarts are famous for their unique orange tails, which are constantly quivered
- Flies and flits about a lot; very shy and nervous

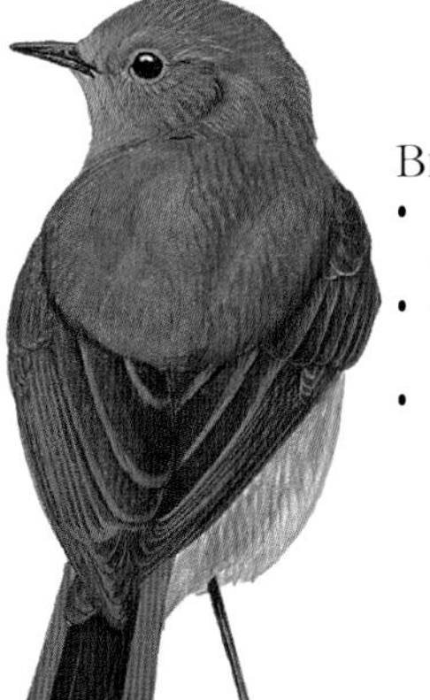

BLACK REDSTART (FEMALE)
- Sooty grey-brown on most of plumage
- Orange tail as in Common - similarly, tail constantly quivered
- Same size and shape as Common

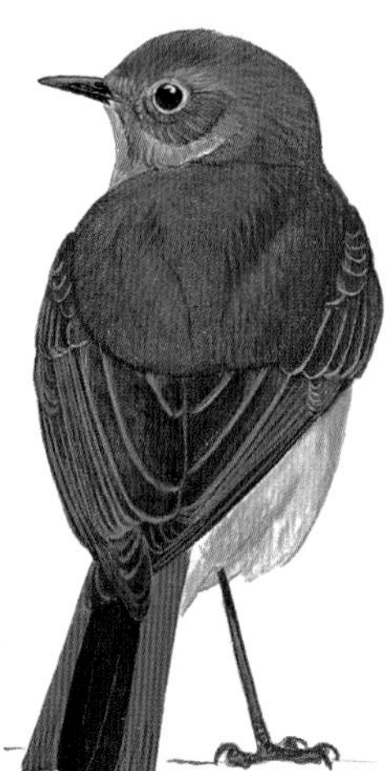

COMMON REDSTART (FEMALE)
- Paler flanks than Black
- Browner upperparts than Black
- More obvious eye-ring than Black

BLACK REDSTART (FEMALE)

- Plain smoky-grey underparts
- Dark throat
- Perches more often on rocks and buildings
  - not usually in bushes or trees

COMMON REDSTART (FEMALE)

- Pale buff underparts; may have hint of orange
- Pale throat

BLACK REDSTART (1ST WINTER)

- Adult males have white secondary panel; on some individuals this may be hinted at early, although bird usually nearly two years old before it is acquired

BLACK REDSTART (1ST-WINTER MALE)

- Not much contrast between colour of upperparts and underparts - considerable contrast in Common
- Sooty brown bird with tail 'set alight'

COMMON REDSTART (1ST-WINTER MALE)

- Clearly defined blackish throat distinguishes it from Common females of all ages, as well as female Black
- Browner crown and mantle than adult's
- Little or no white on forehead
- Reddish-buff edges to flight feathers

COMMON REDSTART (MALE AUTUMN)

- Similar to breeding plumage, but obscured for now by pale fringes to newly acquired feathers (feather tips wear off during winter to reveal striking spring finery)
- Black throat not neatly coloured or defined
- Variable white on forehead
- Flight feathers pale fringed

Common Redstart (female, autumn)

- Breast often slightly reddish (not always); otherwise creamy-buff below
- Greyish-brown above

Black Redstart (female type, autumn)

- Often seen on ploughed fields far from cover; Common most unlikely in such a situation, even on migration

Confusion species

Nightingale (adult, spring)

- Strong, bright rufous-brown on tail and rump (but definitely not orange)
- Bigger than redstarts, and more skulking (does not perch in open)
- Often droops wings and cocks tail (not redstarts)
- Plain face with big, dark eye and pale eye-ring

Bluethroat (1st-winter female)

- Tail has diagnostic reddish sides - hard to miss; unexpectedly bright, and could potentially invite confusion with redstarts if not seen well
- Very long legs - runs on ground (but very skulking - in waterside thickets)
- Frequently cocks tail
- Pale supercilium
- Black spots making ring across breast

STONECHAT & WHINCHAT

INTRODUCTION Whinchat (*Saxicola rubetra*) [L 13cm] widespread summer visitor April–October, to moorland, rough fields, uplands, and so on, often with bracken. European Stonechat (*Saxicola torquata*) [L 12.5cm] resident or short-distance migrant in Britain, Ireland, France, the Low Countries and Germany, with a few outposts elsewhere. Breeds on heaths, coastal scrub, moorland and other open, bushy habitats, often with gorse.

STONECHAT RACES

hibernans (most illustrations) breeds Britain, Ireland and parts of near Continent.
rubicola breeds rest of Europe.

AGEING AND SEXING STONECHAT

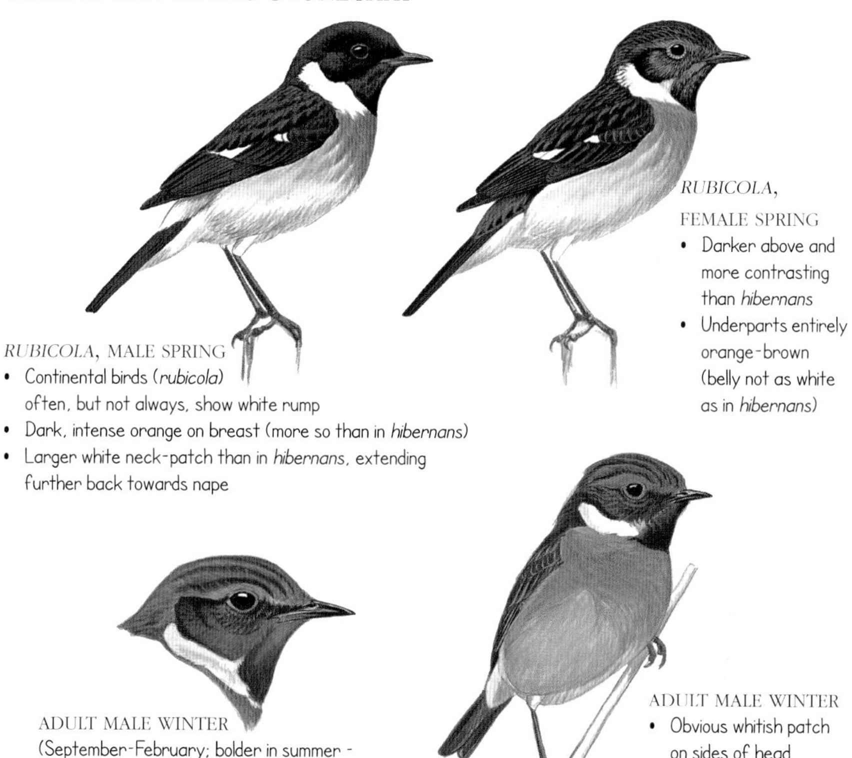

RUBICOLA, FEMALE SPRING

- Darker above and more contrasting than *hibernans*
- Underparts entirely orange-brown (belly not as white as in *hibernans*)

RUBICOLA, MALE SPRING

- Continental birds (*rubicola*) often, but not always, show white rump
- Dark, intense orange on breast (more so than in *hibernans*)
- Larger white neck-patch than in *hibernans*, extending further back towards nape

ADULT MALE WINTER
(September-February; bolder in summer - blacker head and whiter collar)

- Head looks dark at distance
- No obvious supercilium

ADULT MALE WINTER

- Obvious whitish patch on sides of head
- Dark throat
- Warm orange-buff underparts

ADULT MALE WINTER
- Rounded crown (Whinchat has flatter crown)

FEMALE/1ST WINTER
- Pattern as male, but not as bold
- Pale patches on sides of head often smaller than male's

JUVENILE
(June-September only)
- Pale flecks on plumage
- Streaks on breast (c.f. other ages)
- Rich rufous panel on secondaries
- White wing-patch - best giveaway that it is this species

FEMALE
- In all plumages dark tail without extensive white at base
- Darker than Whinchat

DULL FEMALE
- Smaller white patch on wings than male's
- Plain dark tail
- No streaks on rump (c.f. Whinchat)

FEMALE
- Looks a little washed out compared with intricately patterned Whinchat
- Head basically plain

FEMALE/1ST WINTER
(September-February)
- Similar pattern to male's, but paler
- Pale patch on sides of head buff, not white

WHINCHAT

JUVENILE
- No white in wing
- Often a few dark spots on breast-sides
- White tips to scapulars and mantle feathers

JUVENILE
(July-October)
- Bold pale buff supercilium clearest and most consistent field mark
- Whitish throat
- White moustachial stripe separated by narrow dark malar stripe
- Crown streaked with black 'dotted line'

FEMALE
- White base to tail obvious in all plumages
- Small white wing-patches
- Slightly longer and more pointed wings than Stonechat's

JUVENILE
- Paler, more peachy wash to underparts than in Stonechat
- Streaked ear-coverts

WAGTAILS

INTRODUCTION Pied Wagtail (*Motacilla yarrellii*) [L 17.5cm] breeds in Britain and Ireland and the nearby Continent. White Wagtail (*Motacilla alba*) [L 17.5cm] rest of region in a variety of open habitats, including fields, cultivated areas, buildings, waterside and tundra; summer visitor to more extreme climates. Grey Wagtail (*Motacilla cinerea*) [L 18.5cm] breeds widely by rivers, often in hills; summer visitor to Scandinavia. Winters more broadly by water. Yellow Wagtail (*Motacilla flavissima*) [L 15.5cm] summer visitor April–September to Britain and on the North Sea coast from France to southern Norway. Blue-headed Wagtail (*Motacilla flava*) [L 15.5cm] breeds over much of the Continent. Grey-headed Wagtail (*Motacilla thunbergi*) [L 15.5cm] breeds in Scandinavia. All occur in open habitats, especially meadows and grassland, and in some cultivated areas; common migrants.

RARER SPECIES

Spanish Wagtail (*Motacilla iberiae*) [L 15cm] Breeds in south-west France and Iberia; sometimes overshoots northwards in spring.

Black-headed Wagtail (*Motacilla feldegg*) [L 15.5cm] Breeds in south-east Europe. Rare migrant to region.

PIED AND WHITE WAGTAIL ADULTS IN SPRING

PIED WAGTAIL (MALE)

- Black crown and nape, as White
- Mantle and back black, with little or any contrast with nape

WHITE WAGTAIL (MALE)

- Black crown and nape
- Very pale grey back contrasts sharply with black nape, without gradual shading

White Wagtail (female)
• Paler grey upperparts than female Pied's, but slightly darker than male White's
• Black of crown fuses into pale grey of back, with some contrast, but not as sharp as in male
Pied Wagtail (female)
• Dark grey back; not as black as male's, but darker than female White's
• Black of crown and nape fades into dark grey nape, with little contrast
White Wagtail (female)
• Slightly more white on sides of head than in female Pied
Pied Wagtail (female)
• Duskier flanks than female White's

Pied Wagtail (female)

- Dark grey back has diffuse border with black of crown
- Black on rump reaches to tip of shortest tertial

White Wagtail (female)

- Pale grey back has fairly well-defined border with black of crown
- Broader white tertial edges than Pied's
- Rump dark grey

Pied and White Wagtail adults in winter

Pied Wagtail (female autumn)

- Dark grey back has diffuse border with black of crown
- Black on rump reaches to tip of shortest tertial

White Wagtail (female autumn)

- Pale grey back has fairly well-defined border with black of crown
- Broader white tertial edges than Pied's
- Rump dark grey

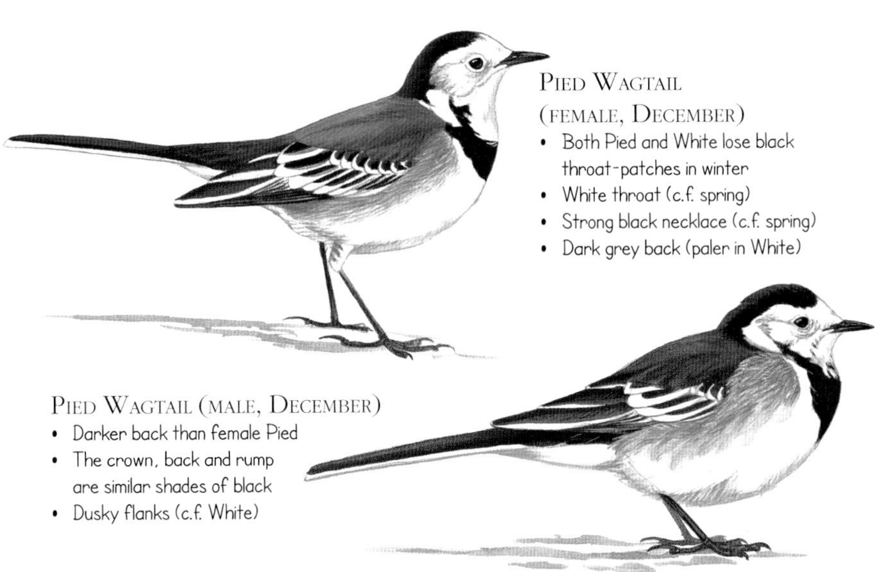

Pied Wagtail (female, December)

- Both Pied and White lose black throat-patches in winter
- White throat (c.f. spring)
- Strong black necklace (c.f. spring)
- Dark grey back (paler in White)

Pied Wagtail (male, December)

- Darker back than female Pied
- The crown, back and rump are similar shades of black
- Dusky flanks (c.f. White)

Juvenile and first winter wagtails

Pied Wagtail (juvenile)

- Greyish cheeks
- Often shows poorly defined breast-band
- Very pale grey crown

Pied Wagtail (early 1st winter)

- Acquires darker crown and clear breast-band
- Hind-crown grey mixed with black

Pied Wagtail (1st winter)

- First-winter Pied and White commonly show yellowish wash on face; ironically, first-winter Yellows do not

White Wagtail (1st winter)

- Very pale crown and back
- Sharply defined dark breast band stands out more starkly
- White forehead

White Wagtail (1st winter)

- Overall neater and cleaner appearance than Pied, especially below, including clean white belly
- Pale grey rump, contrasting with upper tail
- Pale grey flanks, not dusky
- Distinctive ash-grey crown, usually mixed with some black above the supercilium

Pied Wagtail (1st winter)

- Dark-grey flanks and breast-sides
- Smudgy lines on belly (never on White, which is invariably a pale-bellied and extremely smart bird)
- Well defined broad white edges to greater wing coverts
- Much black on head (dark crown and hind-neck), especially contrasting with whitish forehead

Pied Wagtail (juvenile, June)

- Can resemble a White with its grey crown and mantle
- Scruffy and untidy-looking
- Often with strong buff or yellowish wash to face and throat
- Dark and often messy-looking breast-band.

Pied Wagtail (juvenile)

- Much longer tail than Yellow's
- Dusky, and sometimes buff washed on head and flanks
- Dark breast-band or bib
- Broad whitish edges to greater and median wing-coverts (c.f. Yellow)

Yellow Wagtail (juvenile)

- Only yellow is faint tinge on vent (sometimes missing)
- Pale buff supercilium
- Dark necklace around throat (may be indistinct)
- Dark malar stripe

Blue-headed Wagtail (1st winter)

- White supercilium (yellowish in Yellow)
- First-winter birds similar to adult females, but whiter on underparts, with hardly any yellow
- Weak yellow tinge on vent

Blue-headed Wagtail (juvenile)

- Some juveniles very colourless
- Black legs

Adult Yellow and Grey Wagtails

Yellow Wagtail (male)

- More compact than other wagtails
- Indistinct wing-bars
- Undulates strongly, but lacks pronounced sweeping/swooping quality of Grey and Pied/White Wagtails

Grey Wagtail (female)

- Strong, broad white wing-bar both above and below
- Longer tail than Yellow's (and races), giving very different profile
- Strong lemon-yellow patch on vent (on Yellow-type wagtails more buttery yellow)

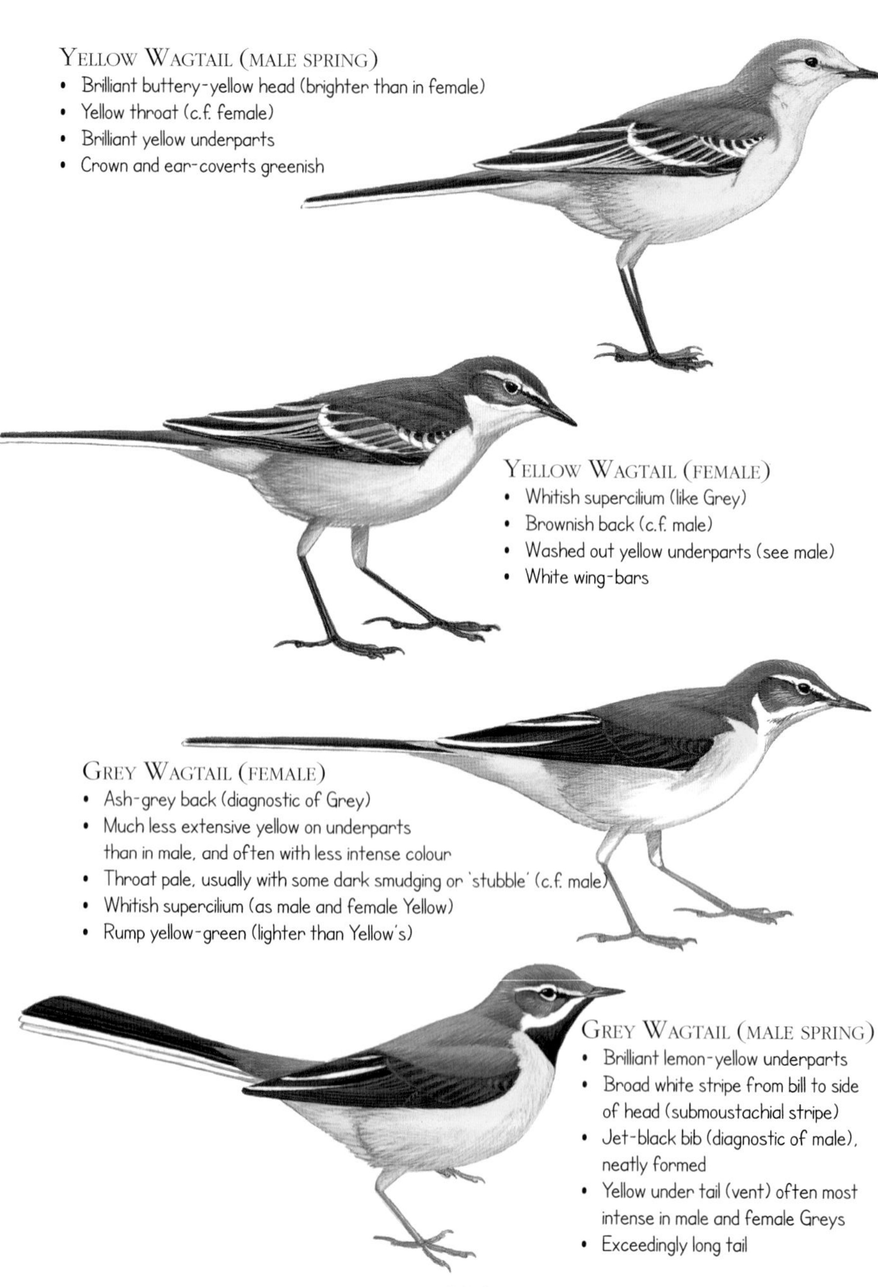
YELLOW WAGTAIL (MALE SPRING)
• Brilliant buttery-yellow head (brighter than in female)
• Yellow throat (c.f. female)
• Brilliant yellow underparts
• Crown and ear-coverts greenish
YELLOW WAGTAIL (FEMALE)
• Whitish supercilium (like Grey)
• Brownish back (c.f. male)
• Washed out yellow underparts (see male)
• White wing-bars
GREY WAGTAIL (FEMALE)
• Ash-grey back (diagnostic of Grey)
• Much less extensive yellow on underparts than in male, and often with less intense colour
• Throat pale, usually with some dark smudging or 'stubble' (c.f. male)
• Whitish supercilium (as male and female Yellow)
• Rump yellow-green (lighter than Yellow's)
GREY WAGTAIL (MALE SPRING)
• Brilliant lemon-yellow underparts
• Broad white stripe from bill to side of head (submoustachial stripe)
• Jet-black bib (diagnostic of male), neatly formed
• Yellow under tail (vent) often most intense in male and female Greys
• Exceedingly long tail

Heads of adult summer 'yellow' wagtails

Yellow Wagtail (male)
- Head mainly yellow
- Yellow supercilium and ear-coverts
- Green eye-stripe and crown

Yellow Wagtail (female)
- Dark ear-coverts
- Greyish crown and nape

Blue-headed Wagtail (female)
- Whitish throat
- Greyer ear-coverts and hind-neck than male's

Blue-headed Wagtail (male)
- Crown and ear-coverts bluish-grey
- White supercilium
- Throat yellow

'Channel' Wagtail (male)
- Hybrid between Blue-headed and Yellow Wagtails
- Noticeably paler head than Blue-headed
- White throat
- White supercilium and pale patch below eye

Spanish Wagtail (male)
- Crown and ear-coverts bluish-grey
- Narrower white supercilium than Blue-headed's, strongest behind eye
- Throat usually white

Grey-headed Wagtail (male)
- Crown sooty-grey (darker than Blue-headed's)
- Ear-coverts dark grey or blackish
- No supercilium (or short one by eye)
- Throat yellow

Black-headed Wagtail (male)
- Whole head glossy black
- No supercilium
- Throat and underparts intense yellow

PIPITS

INTRODUCTION Meadow Pipit (*Anthus pratensis*) [L 14.5cm] widespread on moorland, grassland and pasture, spreading in winter to coasts, estuaries, marshes and farmland. Tree Pipit (*Anthus trivialis*) [L 15cm] widespread summer visitor late March–September, on heaths, woodland edges and clearings; also common passage migrant. Water Pipit (*Anthus spinoletta*) [L 16cm] on upland meadows at 1,400–2,500m. In winter relocates to lowlands, in freshwater habitats such as lake shores and meadows. Rock Pipit (*Anthus petrosus*) [L 16cm] strictly coastal, breeding in northern France, Britain, Ireland and the coast of Scandinavia on rocky shores. In winter disperses, sometimes to fresh water.

Rock Pipit races

petrosus (most illustrations) breeds Britain, Ireland and northern France.
littoralis ('Scandinavian Rock Pipit') breeds Scandinavia, winters to the south, including Britain.

Tree Pipit

- Heavy bill with pink base
- Broad pale supercilium obvious
- Faint but significant dark eye-stripe begins at lores and just about reaches behind eye
- Submoustachial stripe distinctly broad and looks paler than Meadow's
- Often shows pale spot at rear of ear-coverts

Meadow Pipit

- Thinner, sharper bill than Tree's
- Only hint of supercilium
- Plainer face than Tree's, in which pale eye-ring often stands out

Rock Pipit (adult winter)

- Sometimes a hint of a pale supercilium, but white eye-ring quite obvious against dark head
- Upperparts dark grey-brown, lacking warmer tones
- Underparts buff
- Wing-bars buff, not white
- Outer-tail feathers buff (white in Meadow and Tree)

Tree Pipit (adult autumn)

- Streaks on flanks thinner and weaker than those on breast: contrast can be useful pointer
- Hind claws short and strongly curved

Meadow Pipit (adult autumn)

- Streaked evenly down from breast to flanks; streaks just as strong lower down as on breast
- Long hind claws

Tree Pipit (fresh)

- Slightly plumper and more substantial than Meadow
- Bold head markings
- Often shows noticeable contrast between pale buff breast and white belly

Meadow Pipit (fresh)

- Rather plain head, giving slightly gormless look
- No real contrast between ground colour of breast and belly
- Hind-claws long (c.f. Tree)

Pipits in flight

Tree Pipit (display flight)

- Tree Pipit always lifts off from tree to perform song flight, rises steeply, then usually lands on tree when finished (other pipits lift off from ground).

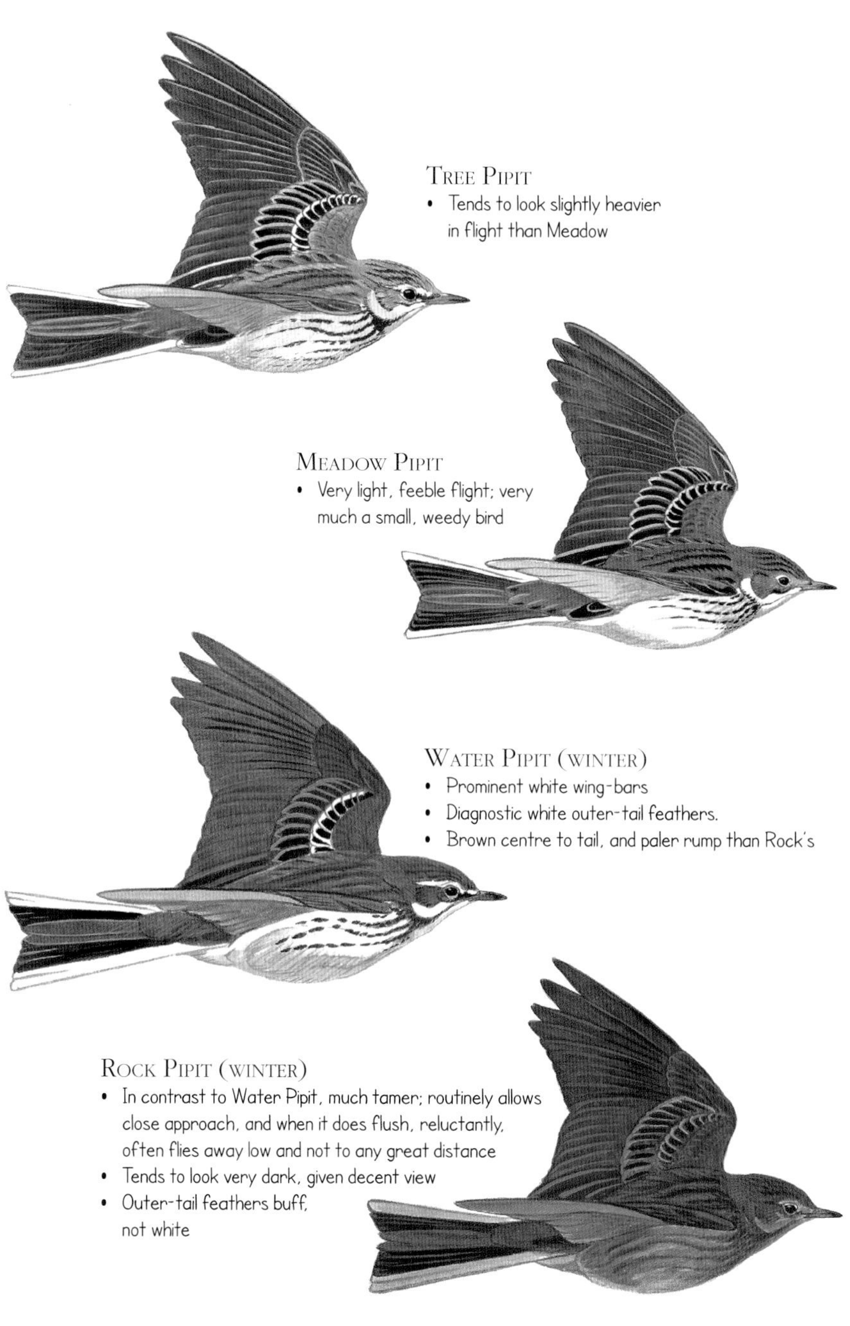
Tree Pipit
• Tends to look slightly heavier in flight than Meadow
Meadow Pipit
• Very light, feeble flight; very much a small, weedy bird
Water Pipit (winter)
• Prominent white wing-bars
• Diagnostic white outer-tail feathers.
• Brown centre to tail, and paler rump than Rock's
Rock Pipit (winter)
• In contrast to Water Pipit, much tamer; routinely allows close approach, and when it does flush, reluctantly, often flies away low and not to any great distance
• Tends to look very dark, given decent view
• Outer-tail feathers buff, not white

Rock Pipit

LITTORALIS, SPRING

- Head with subtle bluish tinge and white supercilium; resembles Water Pipit
- Breast may show subtle tinge of cream, sometimes almost pinkish (as Water), but heavily streaked (= Rock)
- Whitish wing-bars (as Water)
- Dark malar stripe, lacking in Water
- Outer-tail feathers buff, as *petrosus* (some white on very tips)

PETROSUS, SUMMER

- Rather smoky-dark pipit (some greyer than others)
- Very heavy streaking on breast, belly and flanks; streaks somewhat thick and crude
- Ground colour of underparts a sort of 'dirty sheets' creamy-yellow
- Bill black

PETROSUS, WINTER

- Some birds show hint of pale supercilium, but not as obvious as in winter Water, and barely makes it past eye
- Dark head tends to make white eye-ring quite obvious
- Wing-bars buff, not white

PETROSUS, WINTER

- Streaks on breast thick and a little smudged; if they were pencil lines they would be drawn by a 'B' pencil ('H' in Water)
- Ground colour quite dark buff (whiter in Water)
- Rock and Water Pipits both have darker legs (black to very dark red) than the other common pipits (Meadow and Tree)

Water Pipit

SPRING
- Brownish back without obvious streaks; sharp contrast to head
- Pleasing peachy wash to breast
- Sparse streaks on underparts
- Black bill, as Rock (c.f. winter plumage)

AUTUMN, MOULTING
- Base of lower mandible yellow (also throughout winter)
- Greyish hind-neck

WINTER
- Dark legs (c.f. Meadow)
- Subtler streaks on mantle than in Rock
- Whitish underparts with well-defined streaks
- Broad white supercilium

WINTER
- Underparts whiter than Rock's
- Streaks slightly finer, neater and less smudged than Rock's
- Pale supercilium usually well defined and much clearer than in Rock; tends to taper towards rear
- Yellowish bill-base

CHAFFINCH & BRAMBLING

INTRODUCTION Common Chaffinch (*Fringilla coelebs*) [L 15cm] common throughout in all types of woodland, gardens, farmland and hedgerows. Summer visitor to most of Scandinavia; otherwise resident. Brambling (*Fringilla montifringilla*) [L 15cm] northern equivalent, breeding only in Scandinavia, in forests and scrub well into the Arctic. Winters widely south, including Britain.

Chaffinch

FEMALE
- Browner version of male Chaffinch
- Greyish-brown underparts
- Greyish wash behind ear-coverts

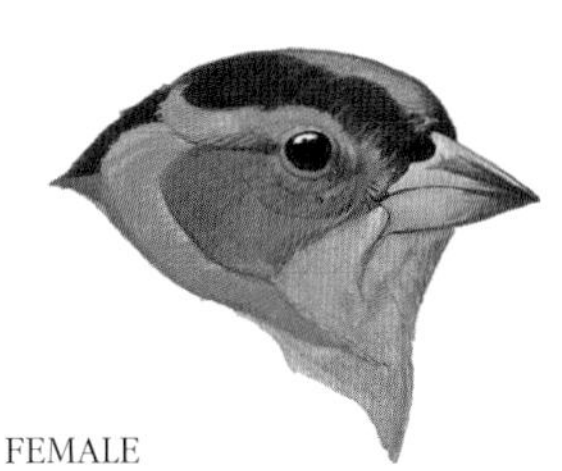

FEMALE
- Dull pink bill with black tip
- Dull brown ear-coverts
- Dark brown sides to crown and pale centre

MALE WINTER
- Reddish-pink breast
- Bluish crown and nape
- Brown back, largely plain
- White shoulders

FEMALE
- White outer-tail feathers
- White shoulders unmistakable

Ageing and sexing Brambling

MALE WINTER
- Orange throat and breast
- Head essentially black, but with orange-buff tips to many feathers giving an untidy look

MALE SPRING
- Smart and unmistakable, with black head and mantle and bright orange breast and scapulars
- Black spots on flanks
- Black bill

1ST-WINTER MALE
- May show less black on head than male

MALE WINTER
- Shorter tail than Chaffinch's, and all black
- Orange breast (similar colour to autumn beech leaves)
- Spots on flanks (never in Chaffinch)
- Orange shoulders

FEMALE WINTER
- Yellow bill with black tip
- Grey frame to ear-coverts, similar to Chaffinch's
- Crown buff-brown with diffuse dark streaks

FEMALE WINTER
- White rump key feature
- Black tail with white only at base of sides

FEMALE WINTER
- Yellow bill (c.f. female Chaffinch)
- Orange throat and breast (c.f. female Chaffinch), but not as intense as in males
- Orange wing-bars (c.f. female Chaffinch)
- Freckles on flanks (c.f. female Chaffinch)

GREENFINCH, SISKIN & SERIN

INTRODUCTION European Greenfinch (*Chloris chloris*) [L 15.5cm] widespread in gardens, open woodland and bushy places. Eurasian Siskin (*Carduelis spinus*) [L 12cm] breeds in extensive coniferous (*especially spruce*) woodland, but winter visitor to lowlands with alder trees. European Serin (*Serinus serinus*) [L 11.5cm] breeds in gardens, parks, cultivated areas and woodland edges on the Continent; mainly summer visitor except resident in south; rare in Britain.

Siskin

MALE
- Coal-black crown (c.f. female)
- Black bib (c.f. female)
- Brilliant yellow-green on head and breast
- Streaked on back (c.f. male Greenfinch)

FEMALE
- Yellow bar goes across wing (c.f. perched Greenfinch)
- Strongly streaked flanks
- Face lacks any prominent markings except pale yellowish supercilium (c.f. male)

FEMALE
- Light, airy, bounding flight
- Two prominent wing-bars
- Streaked yellowish rump (c.f. male)

Lesser Redpoll (female winter)
- Often seen in flocks of Siskins (and vice versa), mixing freely
- Buff, not yellow wing-bars
- Crimson forehead

MALE
- More yellow on tail sides than in female
- Blacker wings than Greenfinch's

Greenfinch

MALE
- No streaks or white on underparts
- Heavier body than Siskin - sparrow-sized

FEMALE
- Large head
- Thick, triangular, pale pink bill
- Yellow panel along edge of wing (not across it, as Siskin)
- Very faint streaks on upperparts and underparts (female only)

MALE
- Black at bill base and around eye, enhancing frowning expression
- Large ash-grey panels on wing
- Grey sides of head
- More yellow on wings and tail than in female

Serin

MALE
- Two narrow whitish wing-bars
- Yellow rump
- Black tail with strong fork

FEMALE
- Compact finch with very distinctive large-headed, yet tiny-billed profile
- Not as yellow as male
- Heavily streaked back

Serin (male)
- Grey, stubby bill key field mark
- Bright lemon-yellow head (c.f. female) and bib
- Ear-coverts dark, surrounded by broad yellow border
- Heavy streaking on white flanks

REDPOLLS

INTRODUCTION Redpolls in their various forms occur in birch woods, heaths, moors, tundra, coniferous woodland and alders. Lesser Redpoll *Carduelis cabaret* [L 12cm] in Britain, Ireland and central Europe; Mealy Redpoll *Carduelis flammea flammea* [13.5cm] in Scandinavia and northern Russia; Arctic Redpoll *Carduelis hornemanni exilipes* (race also known as Coues's Redpoll) [13cm] in Arctic Scandinavia east through Siberia.

MEALY REDPOLL RACE

rostrata (Greenland Redpoll) from southern Greenland and Baffin Island (almost identical to *Carduelis flammea islandica* [Iceland Redpoll] from Iceland).

ARCTIC REDPOLL RACE

hornemanni (race also known as Hornemann's Redpoll) from northern Greenland and Canada (rare vagrant).

LESSER REDPOLL (FEMALE WINTER)

- Distinctly smaller and slimmer than Arctic and Mealy Redpolls
- Always distinctly brown toned, with streaks on underparts set against brown background
- Wing-bars have faint buff tones; never look clean white
- Rump always dark brown, sometimes with a few black streaks; never any hint of white
- Brownish ground colour of breast contrasts with whiter ground colour of belly (feature not shown by the others)

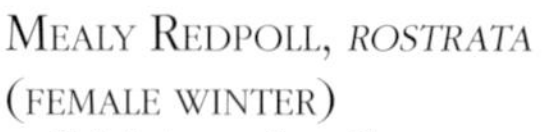

MEALY REDPOLL, *ROSTRATA* (FEMALE WINTER)

- Slightly larger than *flammea*
- Browner head and breast than *flammea*
- Undertail-coverts slightly streaked

Mealy Redpoll, *flammea* (female winter)

- Often distinctive greyish, frosty tinge (as opposed to whitish of Arctic and brownish of Lesser)
- Rump usually has some buff coloration and light streaking, although may look white in flight
- Streaks on back rather strong, giving distinctly striped pattern
- Wing-bar white, but narrower than Arctic's

Arctic Redpoll, *hornemanni* (female winter)

- Larger and whiter than Arctic *exilipes*, including on mantle
- Large head with steep forehead
- Long tail
- Narrow streaks on underparts, especially flanks

Arctic Redpoll, *exilipes* (female type)

- Distinctly chubbier than Mealy, with broader head and neck giving front-heavy look
- Very sparse streaking on breast and flanks
- Slight yellowish tinge to breast-sides
- Broader white wing-bar than in Mealy

Lesser Redpoll (winter)

- Slightly darker red crown than Mealy's
- Head browner than that of others
- Ear-coverts dark

Mealy Redpoll (winter)

- Flashier and more prominent crown than Lesser's (especially) and Arctic's
- Larger bill than Arctic and Lesser's and culmen has curved edge (convex)
- Ear-coverts darker than Arctic's
- Barely discernible supercilium
- Black on nape sometimes more extensive and clearer than in Lesser and Arctic

Arctic Redpoll (winter)

- Bill noticeably smaller than Mealy's; looks pinched in and broad-based
- Culmen of bill straight (see Mealy)
- Distinct creamy, yellowish tinge to head sides and nape, especially in late autumn (often distinctive from Mealy)
- Pale nape with soft, light streaking

Mealy Redpoll

- White wing-bars distinguish it from Lesser
- Usually a few dark streaks on rump; may have unstreaked central area

Arctic Redpoll, *exilipes*

- Usually large area of pure white rump (sugar lump) - extends up to level with tops of tertials
- Broader white fringes to tertials than in Mealy

Mealy Redpoll (undertail)

- Often several dark, arrow-shaped streaks on undertail-coverts; some birds have no streaks

Arctic Redpoll, *exilipes* (undertail)

- Undertail pure white or with one dark streak on longest covert

LINNET & TWITE

INTRODUCTION Common Linnet (*Carduelis cannabina*) [L 12.5cm] widespread in open country, including arable farms, scrub and saltmarshes; summer visitor to southern Scandinavia. Twite (*Carduelis flavirostris*) [L 13cm] the northern equivalent, breeding in Britain, Ireland and Norway, mainly in treeless moorland and grassland, and especially by the coast; winters on saltmarshes, fields and waste places in the North Sea/Baltic and inland Germany.

AGEING TWITE

BREEDING MALE

- Grey bill (same colour as Linnet's) (c.f. non-breeding)
- Pink on rump
- Particularly heavy streaking on body and dark ground-colour

BREEDING FEMALE

- Yellow bill with black tip
- Intense, rich 'cereal-grain' ground colour to head
- Heavily streaked on mantle
- Slightly more diffuse streaks on underparts than in Linnet
- Larger silvery wing-panel

JUVENILE

- Shorter tail than adult's
- Paler throat than adult's

WINTER

- Strongly streaked back and brown appearance recalls Redpoll more than Linnet
- Darker brown on back than Linnet, without any contrast in colour between nape and back
- Deeply forked tail

Twite vs female Linnet in winter

Twite (winter)

LINNET (FEMALE WINTER)
• Throat streaked
• Streaks on breast and flanks peter out less than in Twite
• Different colour on back from Twite: warmer reddish-brown, with fewer streaks
• Strong white edges to tail
• Short grey bill
• Head well patterned, not suffused with one colour as Twite's
• Pale patches above and below eye
• Slim bird, but not quite such a long tail as Twite's
• Even at distance, strongly patterned face still obvious
• Like Twite, gathers into tightly packed, well-coordinated flocks
• Wing pattern less contrasting than Twite's
FEMALE, BREEDING
• Grey bill
• Greyish head with pale patches (e.g. above and below eye)
• Less heavily streaked on back than Twite
• Buff wing-bar

YELLOWHAMMER, CIRL & CORN BUNTINGS

INTRODUCTION Yellowhammer (*Emberiza citrinella*) [L 16cm] widespread in open habitats with bushes and woodland edge. Cirl Bunting (*Emberiza cirlus*) [L 15.5cm] the southern equivalent, mainly in France, Germany and Switzerland, in warm areas; in Britain localized and only in south-west England. Corn Bunting (*Emberiza calandra*) [L 17.5cm] resident in arable crop fields and grassy areas; not in Scandinavia and Ireland. Ortolan Bunting (*Emberiza hortulana*) [L 15.5cm] patchily distributed summer visitor to Scandinavia and parts of the Continent in meadows and bushy country; widespread migrant; rare in Britain.

YELLOWHAMMER

MALE WINTER

- Large, broad, fairly diffuse stripe behind eye (not in female Yellowhammer, and less sharp than in Cirl female)
- Often 'smudged' look between two broad dark stripes on face, making stripes less clean than in Cirl
- Breast marked with heavy, diffuse streaks

FEMALE SUMMER

- Crown usually yellow in middle (not normally in Cirl)
- Nape often has pale patch (not Cirl)
- Rump rich chestnut, without streaks (diagnostic)
- Stripes on breast broader than Cirl's, giving impression of smudged lines rather than sharp pencil lines

JUVENILE

- Often little or no yellow
- Chestnut rump
- Grey bill (c.f. Ortolan)
- Grey sides to head

Ortolan Bunting

JUVENILE

- Pink bill
- Distinctive yellowish eye-ring (unique in region)
- Yellowish throat with dark malar stripe
- Unusual pinkish tinge to underparts, including undertail-coverts

Cirl Bunting

MALE WINTER

- Head often looks a bit more 'sunken in' than Yellowhammer's
- Distinctive face pattern, partly diluted by pale fringes of feathers in winter
- Broad grey-green breast-band (lacking in Yellowhammer)
- Crown dull olive streaked with black

FEMALE

- Strong dark brown stripe begins behind eye, contrasting strongly with pale supercilium
- Face altogether more 'stripy' than female Yellowhammer's
- Breast and flank streaks thinner and finer than Yellowhammer's - clean pencil lines
- Rump dull olive-brown, sometimes with hint of grey, and well streaked (diagnostic)

1ST WINTER

- More stripy-headed than equivalent Yellowhammer
- Olive-grey 'shoulder-patch' not present in Yellowhammer
- Streaks on underparts neater and sharper than any in Yellowhammer

Corn Bunting

- Tail entirely pale brown, lacking any white on sides (contrast Skylark, Reed Bunting)
- Flanks lightly streaked (see Skylark)

- Often dangles legs on short flights (not Skylark, or anything else)

- Much larger and plumper than Yellowhammer or Reed Bunting
- Extremely thick, pale pink bill

- Plump, rounded, heavy body
- Typically, streaks on breast often coalesce to form spot
- Very heavy pink bill

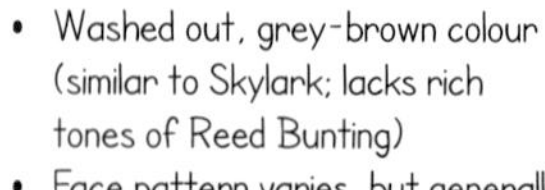

- Washed out, grey-brown colour (similar to Skylark; lacks rich tones of Reed Bunting)
- Face pattern varies, but generally not strong or noticeably stripy
- Back pale brown, lightly streaked (contrast with Reed Bunting)

WINTER

- Bill thick and short (c.f. Skylark)
- Typically seen on wires or in bushes; Skylark avoids both

- Dark malar stripe
- Cheek (ear-coverts) tends to have dark lower border
- Some birds have pale spot on cheek, next to darker one

Corn Bunting confusion species

Skylark (winter)

- Runs along ground; does not hop like Corn Bunting
- Longer, more pointed bill
- Wide pale eye-ring
- Light brown ear-coverts
- Breast streaks concentrated within buff-coloured band across belly (c.f. Corn Bunting)

Reed Bunting (female winter)

- Small greyish bill
- Stripy face completely different from Corn Bunting's
- Warm chestnut fringes to many flight feathers
- Long tail (often twitched), with white outer feathers

REED, LAPLAND & OTHER BUNTINGS

INTRODUCTION Common Reed Bunting (*Emberiza schoeniclus*) [L 14.5cm] widely in marshes and other wetlands, and sometimes drier bushy locations; more varied habitat in winter. Summer visitor Scandinavia. Lapland Bunting (*Calcarius lapponicus*) [L 15cm] common summer visitor to Scandinavia, right up to the Arctic Circle, on tundra where there are bushes and patches of scrub. Winters temperate coasts, on dunes and saltmarshes. Rustic Bunting (*Emberiza rustica*) [L 14cm] breeds sparingly in parts of Scandinavia as a summer visitor; Little Bunting (*Emberiza pusilla*) [L 12.5cm] even more so, mainly in the north-east. Both rare but regular migrants in Britain and elsewhere.

Winter female Lapland and Reed Buntings

Lapland Bunting

- Similar size to Skylark, and often in flocks with it, so easily confused with Skylark in flight
- Strongly notched tail
- In contrast to Reed Bunting, very long wing-tips

Reed Bunting

- Short wings
- Flight style: looks like average small bird, but has a very distinctive flight, with hesitant and jerky style.

Reed Bunting

- Centre of crown pale, but not as pale or prominent as Lapland's
- Prominent pale supercilium
- Ear-coverts darker than supercilium (see Lapland) and bordered with black
- Prominent pale submoustachial stripe (one of most characteristic features)
- Dark malar stripe (also prominent - reaches bill)
- Bill short and dark
- When on ground or perched, distinctive habit of constantly flicking tail

Lapland Bunting

- Dark tail lacks any white borders
- Rather short tail, especially compared with Reed's
- Greater wing-coverts give broad white wing-bar
- Streaks on rump

Reed Bunting

- Prominent white sides to tail
- No white wing-bars; wing fairly uniform rich brown
- Dull brown rump

Lapland Bunting

- Short legs and bulky, but long body
- Noticeably pale centre to crown (contrast Reed)
- Nape buff-coloured at first (early autumn), but in some birds can become quite strongly reddish-brown after mid-winter
- Creamy streaks on back
- Distinctive yellow bill (c.f. Reed Bunting)
- Face relatively plain, with dark, beady eye easily seen
- Ear-coverts fringed with black, widest at corners
- Supercilium and ear-coverts same colour of warm buff, dominating face

Ageing and sexing Reed Bunting

MALE SUMMER

- Very distinctive bold black-and-white head pattern
- Light streaks on whitish breast
- Sings from prominent perch, such as bush or reed top; otherwise skulks

FEMALE SUMMER

- Long pale supercilium
- Signature black malar stripe
- More streaky on breast than male

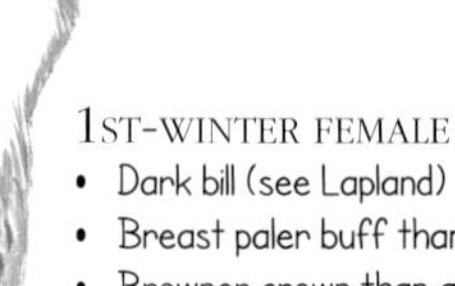

1ST-WINTER FEMALE

- Dark bill (see Lapland)
- Breast paler buff than adult female's
- Browner crown than adult female's
- Usually more heavily streaked on breast than adult

FEMALE WINTER
- Pale brown crown centre
- Pale buff supercilium
- Prominent black malar stripe meets bill base
- Warm buff underparts with extensive streaks
- Ear-coverts grey-brown with darker sides
- Greyish-brown nape

FEMALE SUMMER
- Dark crown
- Dark ear-coverts
- Strong pale supercilium
- Whitish chin

MALE WINTER
- Overall pattern similar to summer male's (see below), but bold colours obscured by buff tips to feathers
- Varying degrees of black blotchiness on throat and upper breast
- Whitish nape

MALE SUMMER
- Bold black head and throat
- Bold white moustachial stripe cuts black in two
- White collar
- Whitish breast

Male and female Lapland Buntings in winter

FEMALE
- Pale central crown-stripe
- Yellow bill with black tip
- Chestnut ear-coverts with black rear edges
- Broad warm buff supercilum

MALE
- Variable black blotches on breast
- Finer streaks on flanks than in Reed

Autumn plumages

Lapland Bunting (female type)
- Bulkier than Reed
- Yellow bill
- Buff face
- Paler buff crown stripe than Reed

Lapland Bunting (male)
- Easily identified by black blotches on breast
- Black legs
- Yellow bill

Little Bunting

- Sharp bill with straight culmen
- Pale lores
- Reddish-brown central crown stripe
- Very black sides to crown
- Chestnut ear-coverts, with dark rear border to ear-coverts
- Pale eye-ring

Rustic Bunting

- Pinkish lower mandible
- Straight culmen (curved in Reed)
- Streaked crown that tends to look peaked
- Brownish nape

Reed Bunting (female)

- Dark grey bill
- Pale brown centre of crown, with light streaks

SNOW BUNTING

INTRODUCTION Snow Bunting (*Plectrophenax nivalis*) [L 17cm] breeds Scotland (rare), Iceland and Scandinavia on tundra and stony mountain tops. Winters widely on temperate beaches, sand-dunes, saltmarshes, mountains and lowland fields.

Ageing and sexing winter Snow Buntings

General features:

- Yellow bill
- Head pattern unusual, with delightful tawny-coloured patches on crown and ear-coverts
- Buff chest-patches
- Black legs
- When feeding, flocks sometimes show continuous 'rolling' action, as birds at back of flock catch up by flying over birds at front
- Extreme examples of age and sex shown here

MALE WINTER

- Shows most amount of white
- Large white wing-panel

1ST-WINTER MALE

- Large white wing-panel intterupted by blackish band across it (on greater wing-coverts)
- Not as white on head as adult male
- More tawny-brown patches on underparts than in adult male
- Primary coverts darker than in adult male

FEMALE WINTER

- More markings on head and nape than in male, e.g. browner ear-coverts
- More brown on underparts than in male
- Rather little white on wings - more bars than panels
- Dark bases to greater wing-coverts

1ST-WINTER FEMALE

- By far brownest of plumages
- Very little white on wing (even, for example, lesser wing-coverts)

ADULT WINTER

- Yellow bill with black tip
- Whitish breast without streaks
- Breast-sides marked with orange-buff, which also meet to form band

MALE WINTER

- Whitish head and face
- Often perches with wings drooped
- Some males have a lot of white on rump (highly variable)

SNOW BUNTING FLOCK

Extremely variable and can be impossible to sex in winter, whether in flight or on ground

- Adult male winter (top right): easy to pick out - mainly white wings with black tips
- 1st-winter female (bottom row, second from left): shows very little white on wing
- 1st-winter males (top row, two leftmost birds): amount of white on inner wing suggests that birds are first-winter males
- Adult female winter (bottom left): has white on wing mainly confined to secondaries, but quite an extensive patch

GLOSSARY

Arm – the inner wing of a bird, nearest the body (before the carpal joint).

Axillaries – the 'armpits of a bird

Bowed – bending downwards, or arched down

Carpal – the area where the front edge of the wing kinks (the 'wrist') is where one finds the carpal joints. Sometimes there are marks here called carpal patches

Chick – general term for a non-flying young bird, usually with fluffy, downy plumage

Coverts – mainly used here to refer to wing coverts, those than make the inner part of the wing behind the primaries and secondaries

Culmen – the upper part (mandible) of bill

Dabbling – working bill at surface of water to acquire food

Diagnostic – especially important for making an identification (diagnosis)

Eclipse – a special plumage worn by certain (usually male) birds to camouflage them when they are moulting their flight feathers

Eye-stripe – a contrasting, usually dark stripe that passes through the eye (see also supercilium)

Fingers – the effect of the primary flight feathers appearing spread and well separated at the end of an open wing, as if they were fingers at the end of the human hand

First winter – a bird living through its first winter of life, having hatched the previous spring/summer

Gape – the fleshy corner or inside of the beak, often brightly coloured

Glide – a type of flight in which the motion is forwards but the wings are not repeatedly flapped

Gonys (Gonydeal Angle) – the ridge along the bottom of a bird's lower mandible. It often angles upwards towards the tip

Graduated – broad at the base but gradually getting narrower towards the end.

Habitat – the type of place in which a bird lives (e.g. marsh, lake, coniferous forest)

Hand – the outer wing of a bird, beyond the carpal joints

Hover – flying 'on the spot', maintaining a steady position in the air by flapping

Immature – any flying stage before adult

Inshore – on the sea close to land (visible from shore)

Juvenile – specifically, a bird that is exhibiting its first full set of feathers, having lost the down of a chick or nestling

Leading edge – the front of the wing as a bird flies

Lore – the area between the bill base and the eye

Morph – consistent different colouration within a species' interbreeding population

Non-breeding – any plumage adopted by a bird for purposes other than breeding (and which is significantly different to breeding plumage)

Offshore – generally not visible from land

Pelagic – habitually occurring far out to sea (usually beyond the Continental Shelf)

Picking – taking an item in the bill directly from the surface

Primary – one of the main flight feathers on a bird (which make up the wing tip)

Primary projection – on the folded wing, the distance the primaries project beyond the covering tertials

Quartering – a slow, forward movement in flight low over the ground

Red-head – a specific term used to denote female/immature diving ducks, especially mergansers

Ringtail – a specific term used to denote female/immature harriers

Secondary – one of a tract of flight feathers that forms the trailing edge of the inner wing

Soar – a type of flight in which the movement is upwards (often in a spiral), but the wings are not repeatedly flapped

Song flight – a ritualised flight following a specific pattern, uttered to the accompaniment of song

Speculum – refers to the usually colourful/iridescent panels on the secondary feathers of ducks

Supercilium – a contrasting stripe (usually light) that passes over the eye (also called eyebrow)

Taiga – a broad habitat zone defined by large areas of coniferous forest

Tail-streamers – tail feathers that are elongated

Trailing edge – the back or hind edge of the wing

Tundra – a broad habitat zone defined by large areas of low-growing, often damp vegetation above the tree line

Vermiculations – closely set wavy lines

Winter plumage – the non-breeding plumage seen on a bird during the winter months

Bibliography

Beaman, M. & Madge, S. 1998. *The Handbook of Bird Identification for Europe and the Western Palearctic*. Christopher Helm, London.

Blomdahl, A., Breife, B. & Holmström, N. 2003. *Flight Identification of European Seabirds*. Christopher Helm, London.

Cramp, S. & Simmons, K.E.L. (eds). 1977–83. *The Birds of the Western Palearctic, Vols 1–3*. OUP, Oxford.

Cramp,S. (ed.). 1985–92. *The Birds of the Western Palearctic, Vols 4–6*. OUP, Oxford.

Cramp, S. & Perrins, C.M. (eds). 1993–94. *The Birds of the Western Palearctic, Vols 7–9*. OUP, Oxford.

Ferguson-Lees, J. & Christie, D.A. 2001. *Raptors of the World*. Christopher Helm, London.

Forsman, D. 1999. *The Raptors of Europe and the Middle East: A Handbook of Field Identification*. T. and A.D. Poyser, London.

Grant, P.J. 1982. *Gulls: A Guide to Identification*. T. and A.D. Poyser, London.

Harris, A., Tucker, L. & Vinicombe, K. 1989. *The Macmillan Field Guide to Bird Identification*. Macmillan, London.

Harris, A., Shirihai, H. & Christie, D.A. 1996. *The Macmillan Birder's Guide to European and Middle Eastern Birds*. Macmillan, London.

Hayman, P. & Hume, R. 2002. *The New Birdwatcher's Pocket Guide to Britain and Europe*. Mitchell Beazley, London.

Madge, S. & Burn, H. 1988. *Wildfowl: An Identification Guide*. Christopher Helm, London.

Olsen, K.M. & Larsson, H. 2004. *Gulls of Europe, Asia and North America*. Christopher Helm, London.

Shirihai, H., Gargallo, G., Helbig, A.J., Harris, A. & Cottridge, D. 2001. *Sylvia Warblers: Identification, Taxonomy and Phylogeny of the Genus Sylvia*. Christopher Helm, London.

Sibley, D. 2000. *The North American Bird Guide*. Chanticleer Press, New York.

Snow, D.W. & Perris, C.M. (eds). 1998. *The Birds of the Western Palearctic*. Concise Edition Vols 1–2. OUP, Oxford.

Svensson, L., Grant, P.J., Mullarney, K. & Zetterström, D. 2011. *Collins Bird Guide, 2nd edition*. HarperCollins, London.

Van Duivendijk, N. 2011. *Advanced Bird ID Handbook: The Western Palearctic*. New Holland, London.

OTHER BOOKS BY NEW HOLLAND

Advanced Bird ID Handbook: The Western Palearctic
Nils van Duivendijk. Award-winning and innovative field guide covering key features of every important plumage of all 1,350 species and subspecies that have ever occurred in Britain, Europe, North Africa and the Middle East. Published in association with the journal *British Birds*.
£24.99 ISBN 978 1 78009 022 1

Bill Oddie's Birds of Britain and Ireland
Bill Oddie. New edition of a best-seller by Britain's best-known birder, and written in Bill's own inimitable style. Covers more than 200 species and ideal for beginners or more experienced birdwatchers.
£12.99 ISBN 978 1 78009 245 4

Birds: Magic Moments
Markus Varesvuo. This award-winning title is a collection of breathtaking photographs that records fleeting moments of drama and beauty in the everyday lives of birds, and allows us all to enter into this exciting and vibrant world.
£20 ISBN 978 1 78009 075 7
(Also available: *Fascinating Birds*
£20 ISBN 978 1 78009 178 5)

The Mating Lives of Birds
James Parry. Bird courtship and display is one of the most spectacular events in the natural world. This beautiful book examines everything from territories and song to displays and raising young.
£19.99 ISBN 978 1 84773 937 7

New Holland Concise Bird Guide
An ideal first field guide to British birds for children or adults. Covers more than 250 species in full colour. Contains more than 800 colour artworks, comes in protective plastic wallet and includes a fold-out insert comparing species in flight. Published in association with The Wildlife Trusts.
£4.99 ISBN 978 1 84773 601 7
Other titles in the *Concise Guide* series (all £4.99): *Butterfly and Moth* (ISBN 978 1 84773 602 4), *Garden Bird* (ISBN 978 1 84773 978 0), *Garden Wildlife* (ISBN 978 1 84773 606 2), *Herb* (ISBN 978 1 84773 976 6), *Insect* (ISBN 978 1 84773 604 8), *Mushroom* (ISBN 978 1 84773 785 4), *Pond Wildlife* (ISBN 978 1 84773 977 3), *Seashore Wildlife* (ISBN 978 1 84773 786 1), *Tree* (ISBN 978 1 84773 605 5) and *Wild Flower* (ISBN 978 1 84773 603 1)

New Holland European Bird Guide
Peter H Barthel. The only truly pocket-sized comprehensive field guide to all the continent's birds. Features more than 1,700 stunning artworks of over 500 species, plus more than 500 distribution maps and a chapter on recognising bird sounds.
£10.99 ISBN 978 1 84773 110 4

Pelagic Birds of the North Atlantic: an ID Guide
Andy Paterson. Innovative new guide, printed on waterproof paper, gives annotated illustrations of every plumage of every pelagic species which could be encountered in the North Atlantic.
£9.99 ISBN 978 1 78009 228 7

Peregrine Falcon
Patrick Stirling-Aird. Beautifully illustrated book detailing the life of this remarkable raptor, including hunting, courtship and raising young. Contains more than 80 stunning colour photographs.
£14.99 ISBN 978 1 84773 769 4
(Also available:
Kingfisher £12.99 ISBN 978 1 84773 524 9;
Barn Owl £14.99 ISBN 978 1 84773 768 7)

The Profit of Birding
Bryan Bland. The author is one of birding's greatest story-tellers, and birders and non-birders alike will enjoy the humorous anecdotal narrative, which is accompanied by many of the author's exquisite line-drawings.
£14.99 ISBN 978 1 78009 124 2

Top 100 Birding Sites of the World
Dominic Couzens. An inspiration for the travelling birder. Brings together a selection of the best places to go birdwatching on Earth, from Norfolk to New Zealand, covering every continent. Includes 350 photos and more than 100 maps.
£19.99 ISBN 978 1 78009 460 1

The Urban Birder
David Lindo. Even the most unpromising cityscapes can be great for birds. Includes tales of gun-toting youths and migration-watching from skyscrapers.
£9.99 ISBN 978 1 84773 950 6

See www.newhollandpublishers.com for further details and special offers

INDEX

Bold = main section on species